Lecture Notes in Mathematics

Edited by A. Dold and B. Eckmann

Series: Australian National University, Canberra

Advisers: L. G. Kovács, B. H. Neumann and M. F. Newman

697

Topics in Algebra

Proceedings, 18th Summer Research Institute
of the Australian Mathematical Society
Australian National University
Canberra, January 9 – February 17, 1978

Edited by M. F. Newman

Springer-Verlag
Berlin Heidelberg New York 1978

Editor
M. F. Newman
Department of Mathematics
Institute of Advanced Studies
Australian National University
PO Box 4, Canberra 2600
ACT
Australia

Assistant Editor
J. S. Richardson
Department of Mathematics
Institute of Advanced Studies
Australian National University
PO Box 4, Canberra 2600
ACT
Australia

AMS Subject Classification (1970): 13-01, (13B25), 15-01 (15A36), 16-02, (16A38), (16A68), 17B10, (17B60), 20-02, 20C10, 20C15, (20C30), 20D99, 20E10, (20F05), 20F35, 20F40, (20F50), 20K20, (20K40), (20K99)

ISBN 3-540-09103-3 Springer-Verlag Berlin Heidelberg New York
ISBN 0-387-09103-3 Springer-Verlag New York Heidelberg Berlin

Printing and binding: Beltz Offsetdruck, Hemsbach/Bergstr.
2141/3140-543210

INTRODUCTION

The Summer Research Institute of the Australian Mathematical Society is an annual event held each (southern hemisphere) summer with the aim of providing stimulus and opportunity for research in mathematics. The main theme (as suggested by the Society) for the 18th SRI (1978) was algebra. A number of leading algebraists from overseas, with diverse interests within algebra, were invited to participate and help formulate a view of current directions in research in algebra and the role of algebra and algebraic research both within mathematics and as related to other disciplines. As well a number of other leading overseas and Australian mathematicians were invited to contribute lectures or organize and participate in splinter groups. Details are given in an appendix to this introduction.

The participants in the SRI seem to have found the lectures of good value; their enthusiasm was sufficient to encourage us to offer texts to a wider audience in this form.

These proceedings consist, in the main, of those invited lectures and splinter group talks in algebra which are not otherwise available (and for which manuscripts were received). Thus Professor Irving Kaplansky's six-lecture survey of "Algebraic aspects of Hilbert's problems" is not here; much of the mathematical information presented, and a good deal more, can be found in his "Hilbert's Problems, Preliminary Edition" (1977) which occurs in the series: Lecture Notes in Mathematics, of the Department of Mathematics, University of Chicago. Professor Jacobson provides, in the introduction to his lectures in these proceedings, references to the other material he presented.

Comparison of lectures and their impact is a notoriously subjective and, therefore, invidious matter. Nevertheless I will venture a small step in that direction by suggesting that the message of Sims' lecture "The role of algorithms in the teaching of algebra" met with considerable support.

I am much indebted to the other officers of the SRI, R.A. Bryce, G. Havas and L.G. Kovács for their strong support in organizing the SRI and to J.S. Richardson for his work as Assistant Editor of these proceedings.

M.F. Newman,
Director

LIST OF PARTICIPANTS

G.Q. Abbasi, Australian National University, Canberra, ACT.
M. Adelman, Macquarie University, North Ryde, New South Wales.
R.S. Anderssen, Australian National University, Canberra, ACT.
J.A. Ascione, Australian National University, Canberra, ACT.
A. Baker, University of Cambridge, Trinity College, Cambridge, England.
D. Barnes, University of Sydney, Sydney, New South Wales.
P.M. Beazley, University of Sydney, Sydney, New South Wales.
T. Bisson, Duke University, Durham, North Carolina, USA.
T.G. Brook, Australian National University, Canberra, ACT.
M.S. Brooks, Canberra College of Advanced Education, Canberra, ACT.
B.M. Brown, La Trobe University, Bundoora, Victoria.
R.A. Bryce, Australian National University, Canberra, ACT.
G. Butler, University of Sydney, Sydney, New South Wales.
R.N. Buttsworth, University of Queensland, St Lucia, Queensland.
J.M. Campbell, Canberra College of Advanced Education, Canberra, ACT.
M.J. Canfell, University of New England, Armidale, New South Wales.
J.J. Cannon, University of Sydney, Sydney, New South Wales.
G.A. Chandler, Australian National University, Canberra, ACT.
J.H. Coates, Australian National University, Canberra, ACT.
S.B. Conlon, University of Sydney, Sydney, New South Wales.
C.D.H. Cooper, Macquarie University, North Ryde, New South Wales.
W.A. Coppel, Australian National University, Canberra, ACT.
P.J. Cossey, Australian National University, Canberra, ACT.
J. Cresp, Australian National University, Canberra, ACT.
J.N. Crossley, Monash University, Clayton, Victoria.
F.R. de Hoog, CSIRO, Canberra, ACT.
P. Donovan, University of New South Wales, Kensington, New South Wales.
M.W. Evans, Department of Education, Melbourne, Victoria.
M.J. Field, University of Sydney, Sydney, New South Wales.
B. Fischer, University of Bielefeld, Bielefeld, West Germany.
P. Fitzpatrick, Australian National University, Canberra, ACT.
T.M. Gagen, University of Sydney, Sydney, New South Wales.
J.P. Glass, Sydney, New South Wales.
G.H. Golub, Stanford University, Stanford, California, USA.
A.O. Griewank, Australian National University, Canberra, ACT.
G.E. Griffiths, University of New South Wales, Kensington, New South Wales.
J.R.J. Groves, University of Melbourne, Parkville, Victoria.
S.A. Gustavson, Australian National University, Canberra, ACT.
W. Haebich, National Mutual Assurance Company, Melbourne, Victoria.
G. Havas, Australian National University, Canberra, ACT.
J.L. Hickman, Australian National University, Canberra, ACT.
J.L. Hillman, Australian National University, Canberra, ACT.
K. Horadam, Murdoch University, Murdoch, Western Australia.
T.S. Horner, University of Wollongong, Wollongong, New South Wales.
G.A. How, Australian National University, Canberra, ACT.
R. Howlett, University of Sydney, Sydney, New South Wales.
D.C. Hunt, University of New South Wales, Kensington, New South Wales.
F.D. Jacobson, Albertus Magnus College, Connecticut, USA.
N. Jacobson, Yale University, New Haven, Connecticut, USA.
P. Johnson, University of Sydney, Sydney, New South Wales.
I. Kaplansky, University of Chicago, Chicago, Illinois, USA.

G.M. Kelly, University of Sydney, Sydney, New South Wales
B.W. King, Riverina College of Advanced Education, Wagga Wagga, New South Wales.
L.G. Kovács, Australian National University, Canberra, ACT.
R.M. Kuhn, University of Sydney, Sydney, New South Wales.
P. Laird, University of Wollongong, Wollongong, New South Wales.
M. Lazard, University of Paris, Paris, France.
G. Lehrer, University of Sydney, Sydney, New South Wales.
R.H. Levingston, University of Sydney, Sydney, New South Wales.
R. Lidl, University of Tasmania, Hobart, Tasmania.
J. Loxton, University of New South Wales, Kensington, New South Wales.
J. Mack, University of Sydney, Sydney, New South Wales.
W.G. McCallum, University of New South Wales, Kensington, New South Wales.
D.J. McCaughan, University of Otago, Dunedin, New Zealand.
A. McIntosh, Macquarie University, North Ryde, New South Wales.
J.A. McPhee, Royal Melbourne Institute of Technology, Melbourne, Victoria.
K. Mahler, Australian National University, Canberra, ACT.
P. Majstrenko, University of New England, Armidale, New South Wales.
C.F. Miller, University of Melbourne, Parkville, Victoria.
B. Musidlak, University of New South Wales, Kensington, New South Wales.
B.H. Neumann, Australian National University, Canberra, ACT.
M.F. Newman, Australian National University, Canberra, ACT.
S. Okada, Australian National University, Canberra, ACT.
M.R. Osborne, Australian National University, Canberra, ACT.
J. Paine, Australian National University, Canberra, ACT.
K.J. Palmer, Australian National University, Canberra, ACT.
K.R. Pearson, La Trobe University, Bundoora, Victoria.
C.D. Pischettola, University of Adelaide, Adelaide, South Australia.
J. Pitman, University of Adelaide, Adelaide, South Australia.
B. Ponnudurai, University of Adelaide, Adelaide, South Australia.
I. Reiner, University of Illinois, Urbana, Illinois, USA.
J.C. Renaud, University Papua New Guinea, Papua New Guinea.
J.S. Richardson, Australian National University, Canberra, ACT.
R.W. Richardson, Australian National University, Canberra, ACT.
D.K. Ross, La Trobe University, Bundoora, Victoria.
J.H. Rubinstein, University of Melbourne, Parkville, Victoria.
M.G. Schooneveldt, Australian National University, Canberra, ACT.
P. Schultz, University of Western Australian, Nedlands, Western Australia.
D.B. Shield, Goroko Teachers College, Goroka, Papua New Guinea.
C.C. Sims, Rutgers University, New Brunswick, New Jersey, USA.
P.D. Smith, Australian National University, Canberra, ACT.
R.G. Stanton, University of Manitoba, Winnipeg, Manitoba, Canada.
L.S. Sterling, Australian National University, Canberra, ACT.
A.H. Stone, University of Rochester, Rochester, New York, USA.
D.M. Stone, University of Rochester, Rochester, New York, USA.
R. Street, Macquarie University, North Ryde, New South Wales.
E. Szekeres, Macquarie University, North Ryde, New South Wales.
G. Szekeres, University of New South Wales, Kensington, New South Wales.
D.E. Taylor, University of Sydney, Sydney, New South Wales.
R.L. Tweedie, CSIRO, Canberra, ACT.
C.J.F. Upton, University of Melbourne, Parkville, Victoria.
A.J. van der Poorten, University of New South Wales, Kensington, New South Wales
M. Waldschmidt, University of Paris, Paris, France.
G.E. Wall, University of Sydney, Sydney, New South Wales.
J.A. Ward, Australian National University, Canberra, ACT.
J.N. Ward, University of Sydney, Sydney, New South Wales.
R.F.C. Walters, University of Sydney, Sydney, New South Wales.
K.S. Watson, Flinders University, Bedford Park, South Australia.
J.B. Wilker, University of Toronto, Toronto, Ontario, Canada.
W.H. Wilson, University of Queensland, St Lucia, Queensland.
R. Worley, Monash University, Clayton, Victoria.
R.L. Yager, Australian National University, Canberra, ACT.
S. Yamamuro, Australian National University, Canberra, ACT.

INVITED GENERAL LECTURES
(in chronological order)

I. Reiner, "Integral representations: genus, K-theory, and class groups" (3 lectures).

A. Baker, "Transcendence theory and its applications".

I. Kaplansky, "Algebraic aspects of Hilbert's problems" (6 lectures).

G.H. Golub, "Developments and trends in numerical linear algebra".

C.C. Sims, "The role algorithms have played and should play in algebraic research and instruction".

M. Lazard, "Wer sind und sollen Evariste Galois und Richard Dedekind - a historical approach to contemporary algebra".

B. Fischer, "Involutory generators for finite groups".

G.H. Golub, "The inverse eigenvalue problem".

M. Waldschmidt, "Transcendental numbers, functions of several variables, and algebraic hypersurfaces".

N. Jacobson, "The development of the theory of Jordan algebras".

G.E. Wall, "Commutator calculus and Lie methods in groups" (2 lectures).

C.C. Sims, "Some group theoretic algorithms".

W.A. Coppel, "Algebra from control theory".

B. Fischer, "Subgroups of finite simple groups".

J.H. Coates, "Iwasawa's theory of $\mathbb{Z}_p$-extensions of number fields".

R.W. Richardson, "Commuting pairs of elements in semi-simple Lie algebras".

N. Jacobson, "The history of algebras with polynomial identity".

N. Jacobson, "Some recent advances in the theory of algebras with polynomial identity" (4 lectures).

I. Kaplansky, "Lie and Jordan superalgebras".

M. Rabin (on videotape), "Probalistic test for primality".

M. Rabin (on videotape), "Algorithms in finite fields and number theory".

SPLINTER GROUPS
(in chronological order)

REPRESENTATIONS OF GROUPS AND ALGEBRAS, organized by S.B. Conlon.

I. Reiner, "Integral representations of cyclic groups of prime-power order" (3 lectures).

G. Lehrer, "Regular representations of algebraic groups over finite fields".

R. Howlett, "Characters of groups with normal extraspecial subgroups".

P. Donovan, "Dihedral defect".

L.G. Kovács, "Some representations of general linear groups".

NUMBER THEORY, organized by J.H. Coates

K. Mahler, "A 'theorem' on continuity of p-adic functions".

J.P. Glass, "Theta functions of genus 3 ".

P. Majstrenko, "Applications of finite algebraic identities to rational points on elliptic curves".

A. Baker, "Diophantine analysis".

A.J. van der Poorten, "Elliptic curves of conductor eleven".

M. Waldschmidt, "Transcendental numbers and group varieties".

J.H. Coates, "P-adic L-functions and elliptic units".

A.J. van der Poorten, "Transcendence properties of a class of transcendental functions first studied by Mahler".

NUMERICAL LINEAR ALGEBRA, organized by R.S. Anderssen and M.R. Osborne.

L.S. Sterling, "Canonical forms for matrices related to abelian groups".

J.A. McPhee, "Computation of eigenvectors".

G.H. Golub, "Computing singular values".

J. Paine, "Eigenvalues for Sturm-Liouville problems".

S.A. Gustavson, "Calculating linear functions".

G.H. Golub, "Error bounds for linear systems using the theory of moments".

G.A. Chandler, "Approximate solution of operator equations".

M.R. Osborne, "Stabilized marching technique".

R.S. Anderssen, "Differential inverse eigenvalue problems".

F.R. de Hoog, "Asymptotic expansions for boundary value problems".

ALGEBRAIC ALGORITHMS AND THEIR IMPLEMENTATION, organized by J.J. Cannon.

C.C. Sims, "Permutation group algorithms" (4 lectures).

B. Fischer, "Character tables".

G. Havas, "Computational results related to the Burnside problem".

G. Butler, "Computing conjugacy classes in permutation groups".

J.S. Richardson, "Simplifying group presentations".

Problem Session.

M.F. Newman, "Classifying p-groups".

FINITE INSOLUBLE GROUPS, organized by T.M. Gagen and D.C. Hunt.

R. Howlett, "Removing the N from BN".

B. Fixcher, "Character tables of group extensions".

R.H. Levingston, "Permutation groups containing a cycle of prime-power length".

G. Lehrer, "Representations of algebraic groups over finite fields".

CATEGORY THEORY, organized by R. Street.

G.M. Kelly, "Traces and categories whose objects are small projective".

J. Flood, "Semi-convex geometry".

R. Street, "Continuously variable categories" (2 lectures).

COMMUTATOR CALCULUS AND LIE METHODS IN GROUPS, organized by G.E. Wall.

D.B. Shield, "Power-commutator structure of p-groups".

M. Lazard, "The formula for $\log(e^x e^y)$".

G.E. Wall, "Finite groups of prime exponent" (2 lectures).

G.E. Wall, "Commutator collection preserving module structure".

L.G. Kovács, "Kljačko's work on varieties of nilpotent groups of small class" (2 lectures).

G.E. Wall led a small but intensive study group reading Kostrikin's work on the restricted Burnside problem (6 meetings).

PROBABILITY, organized by R.L. Tweedie.

R.L. Tweedie, "Markov chain transition probabilities: connections between decompositions, orthogonalities and topological properties".

D.M. Stone, "An example of orthogonality in a family of probability measures".

RINGS AND ALGEBRAS, organized by D.W. Barnes.

W.H. Wilson, "Induced representations of Lie algebras".

I. Kaplansky, "Trilinear forms".

K. Watson, "Manifolds over Jordan pairs".

W.H. Wilson, "Clifford's Theorem out of context".

D.W. Barnes, "Saturated formations of soluble Lie algebras in characteristic zero".

INFINITE GROUPS AND LOGIC, organized by C.F. Miller.

J.R.J. Groves, "Ultraproducts and groups".

C.F. Miller, "Modules, algorithms, and embeddings for groups".

J.R.J. Groves, "Finitely presented soluble groups".

C.F. Miller, "Small cancellation theory and the method of diagrams" (3 lectures).

C.F. Miller, "Work of Baumslag, Dyer, and Heller on acyclic groups".

J.H. Rubinstein, "Fundamental groups of 3-manifolds" (2 lectures).

ALGEBRA WITHOUT THE AXIOM OF CHOICE, organized by J.L. Hickman.

B.H. Neumann, "Semigroups".

J.L. Hickman, "Lattices".

AD HOC LECTURES

W. Haebich, "The cohomology of certain products of groups, without spectral sequences".

I. Kaplansky, "An informal survey of developments in the theory of abelian groups".

M.J. Field, "Differentiable invariants of Lie groups".

CONTENTS

PROC. 18th SRI,
CANBERRA 1978, 1-7.

20C15

EXTENDING CHARACTERS FROM NORMAL SUBGROUPS

Robert B. Howlett

The purpose of this note is to give a short proof of the main theorem of [1]. Essentially the same simplification has also been discovered (independently) by Isaacs. The theorem is

THEOREM 1. *Suppose that* A *is a finite group which acts on the finite group* H *, and*

(1) *for some prime* r *,* $|H'| = r$ *and* $H/Z(H)$ *is an elementary abelian* r*-group,*

(2) A *centralizes* $Z(H)$ *,*

(3) A *has a soluble normal subgroup* B *with order prime to* r *and satisfying* $[H, B]Z(H) = H$ *.*

Then any non-linear irreducible complex character of H *can be extended to a character of* AH *.*

We treat even and odd r simultaneously, although a short proof of a different kind is available for odd r (see [1]). One application of the theorem is the proof for soluble groups G of the following (McKay's conjecture): if N is the normalizer of a Sylow p-subgroup of G , the number of irreducible complex characters with degree prime to p is the same for G as for N . This is proved in a paper by Wolf [6].

Before starting the main part of the proof we collect into a lemma three well known sufficient conditions for the extendibility of a character.

LEMMA 2. *Suppose* $H \triangleleft G$, $G = TH$, $T \cap H = 1$. *Let* χ *be an irreducible complex character of* H *such that for all* $t \in T$, $h \in H$, $\chi(t^{-1}ht) = \chi(h)$. *Then* χ *can be extended to* G *if any of the following hold:*

(a) T *is cyclic;*

(b) $\chi(1)$ *and* $|T|$ *are coprime;*

(c) for each prime q *dividing* $|T|$ *there exists a Sylow* q*-subgroup* Q *of* T *such that* χ *extends to* QH.

Proof. Part *(a)* is easy. For part *(b)* see [3]. (Note that *(b)* includes the special case $\chi(1) = 1$.) For part *(c)* see [5].

LEMMA 3. *Let* $H \triangleleft G$, χ *an irreducible character of* H, *and*

$$K = \{g \in G \mid \chi(g^{-1}hg) = \chi(h) \text{ for all } h \in H\} .$$

Then if ϕ *is any irreducible component of the induced character* χ^K *then* ϕ^G *is irreducible.*

Proof. See [2, Theorem 1].

Now let A and H satisfy the hypotheses of Theorem 1. The proof proceeds by induction on $|AH|$. We may regard $H^* = H/Z(H)$ as a vector space over F_r, the field with r elements. We use additive notation in H^* and use stars to denote images in H^* of subgroups of H. Conjugation of elements of H by elements of A induces an action of A on H^*, making H^* into an F_rA-module. As an F_rB-module H^* is completely reducible (by Maschke's Theorem).

PROPOSITION 4. *Suppose that* $Z(H) \leq M \leq H$ *and that* M *is* B*-invariant. Then* $[M, B]Z(H) = M$. *If* M *is abelian* $M = [M, B] \times Z(H)$.

Proof. Let N be a subgroup of H with $H^* = M^* \oplus N^*$ and N B-invariant. Since $[H, B]^* = H^*$, clearly $[M, B]^* = M^*$ and $[N, B]^* = N^*$, proving the first assertion. The second assertion follows from [4, Theorem 5.2.3].

PROPOSITION 5. *Let* K *be any subgroup of* H *such that* $Z(K) = Z(H)$. *Then* H *is the central product of* K *and* $C_H(K)$ *with* $Z(K) = Z(C_H(K))$ *amalgamated.*

Proof. See [4, Lemma 5.4.6].

It is easily shown that for any linear character λ of $Z(H)$ which is non-trivial on H', λ^H has a unique irreducible constituent χ which satisfies

$$\chi(x) = 0 \qquad \text{if } x \notin Z(H) ,$$

$$\chi(x) = m\lambda(x) \quad \text{if } x \in Z(H) ,$$

where $m = \chi(1)$ is the multiplicity of χ in λ^G, and $m^2 = [H : Z(H)]$. Each non-linear irreducible character of H is obtainable in this way. For the rest of the proof χ and λ will be fixed, and we assume that χ does not extend to AH. (Thus A, H and χ constitute a minimal counterexample.) Minimality of A implies that A acts faithfully on H.

PROPOSITION 6. *H^* is an irreducible F_rA-module.*

Proof. Suppose not, and let M^* be an irreducible F_rA-submodule of H^*.

Case (i). Suppose that M is non-abelian. Then $Z(M) = Z(H)$ and so $H = K_1K_2$ with $K_1 = M$, $K_2 = C_H(M)$. The K_i are A-invariant, and using Proposition 4 we see that the hypotheses of the theorem are satisfied when K_i replaces H. So we will be able to apply induction.

By the representation theory of central products there exist irreducible $\mathbb{C}K_i$-modules V_i $(i = 1, 2)$ such that under the action

$$x_1x_2(v_1 \otimes v_2) = x_1v_1 \otimes x_2v_2 \quad (x_i \in K_i, v_i \in V_i)$$

$V_1 \otimes V_2$ is a well defined $\mathbb{C}H$-module affording the character χ. By induction V_i can be made into a $\mathbb{C}AK_i$-module. Now if we define

$$a(v_1 \otimes v_2) = av_1 \otimes av_2 \quad (a \in A, v_i \in V_i)$$

then $V_1 \otimes V_2$ becomes a $\mathbb{C}AH$-module, and its character extends χ - a contradiction.

Case (ii). Suppose that M is abelian, and let N be a maximal abelian A-invariant subgroup of H containing M. Then $N = [B, N] \times Z(H)$. Define μ to coincide with λ on $Z(H)$ and have kernel L containing $[B, N]$. Then μ is fixed by A, and so the inertia group $K = \{x \in H \mid \mu^x = \mu\}$ is A-invariant. Let ψ be an irreducible constituent of μ^K. By Lemma 3, ψ^H is irreducible, and since it is a constituent of λ^H, $\psi^H = \chi$.

Suppose firstly that $K = N$. By Lemma 2 *(b)*, $\psi = \mu$ extends to a linear character $\tilde{\psi}$ of AK, and obviously $\tilde{\psi}^{AH}$ extends $\psi^H = \chi$ - a contradiction. Suppose on the other hand that K is non-abelian. Since $L \lhd K$ we may set $\overline{K} = K/L$. Because $L \cap K' = 1$, $|\overline{K}'| = |K'| = r$. Moreover, $Z(\overline{K}) = Z(K)L/L = N/L = Z(H)L/L$, and so A centralizes $Z(\overline{K})$. Now we may apply induction (with $A, \overline{K}, \mu, \psi$ replacing A, H, λ, χ) to conclude that ψ extends to a character $\tilde{\psi}$ of AK. As before, $\tilde{\psi}^{AH}$ extends ψ^H, a contradiction.

PROPOSITION 7. *If L is a Sylow r-subgroup of A then $A = LB$.*

Proof. If $LB \neq A$ then by induction χ extends to LBH. If q is any other prime dividing $|A|$ and Q a Sylow q-subgroup of A then by Lemma 2 *(b)*, χ extends to QH. By Lemma 2 *(c)* it follows that χ extends to AH, a contradiction.

PROPOSITION 8. *B is a minimal normal subgroup of A.*

Proof. Suppose not, and let B_0 be a minimal normal subgroup of A contained in B. Let $A_0 = LB_0$, so that $|A_0| < |A|$. Since it suffices to prove that χ extends to LH it suffices to prove that χ extends to A_0H. If $[B_0, H]Z(H) = H$ this follows by induction since all the hypotheses of the theorem are satisfied when A_0 replaces A.

If $[B_0, H]Z(H) < H$ then by Proposition 6, $[B_0, H] \leq Z(H)$. By [4, Theorem 5.3.5], $C_H(B_0) = H$. But this is impossible since B_0 is a nontrivial group of automorphisms of H. Hence $[B_0, H]Z(H) = H$, as required.

In view of Proposition 6 we may apply Clifford's Theorem [4, Theorem 3.4.1] and write H^* as an F_rB-module direct sum

$$H^* = H_1^* \oplus H_2^* \oplus \ldots \oplus H_n^*$$

where the H_i are permuted transitively by L. The H_i^* are the F_rB-primary components of H^* (that is, H_i^* is the sum of all the irreducible F_rB-submodules of H isomorphic to a given irreducible module).

PROPOSITION 9. *If B is not cyclic there exist B-invariant proper subgroups $K_1, K_2, \ldots, K_m$ of H such that*

(1) *$K_1, K_2, \ldots, K_m$ are permuted transitively by L,*

(2) *H is the central product $K_1K_2 \ldots K_m$ with*

$$Z(K_1) = Z(K_2) = \ldots = Z(K_m) = Z(H)$$

amalgamated.

Proof. By Proposition 8 and the fact that B is soluble, B is abelian. Assuming that B is not cyclic it follows that the subgroups $C_H(x)$ $(x \in B)$ generate H [4, Theorem 5.3.16]. Let B_1 be a maximal subgroup of B such that $K_1 = C_H(B_1) \neq Z(H)$. If M^* is any F_rB-submodule of H^* isomorphic to some submodule of K_1^* then B_1 acts trivially on M^*, and by [4, Theorem 5.3.15], $M^* \subseteq K_1^*$. Thus K_1^* is a direct sum of some subset of the primary components H_i^*. Since

$$H^* = K_1^* \oplus [B_1, H]^* \quad ([4, \text{Theorem } 5.2.3])$$

it follows that $[B_1, H]^*$ is the sum of the remaining primary components. Now since $K_1 \lhd H$ and $[K_1, B_1] = 1$,

$$[K_1, B_1, H] = [H, K_1, B_1] = 1$$

and so by the three subgroup lemma ([4, Theorem 2.2.3]), $[B_1, H]$ and K_1 centralize each other.

Let $l \in L$, $l \notin N_L(B_1)$. Then

$$C_H(B_1) \cap C_H\left(B_1^l\right) = C_H\left(B_1B_1^l\right) = Z(H)$$

by choice of B_1. Hence $\left(K_1^l\right)^*$ is a sum of primary components distinct from those in K_1^*. So K_1 and K_1^l centralize each other. Similarly, if T is a set of representatives of the right cosets of $N_L(B_1)$ in L then distinct members of

$$\left\{\left(K_1^l\right)^* \mid l \in T\right\} = \{K_1^*, K_2^*, \ldots, K_m^*\}$$

do not have a primary component in common. Hence their sum is direct. Furthermore, $K_1, K_2, \ldots, K_m$ centralize each other. Since L permutes $K_1, K_2, \ldots, K_m$ it follows that $K_1K_2 \ldots K_m$ is A-invariant. Hence $H = K_1K_2 \ldots K_m$, as required.

PROPOSITION 10. *B is cyclic of prime order p.*

Proof. By Proposition 8, B is an elementary abelian p-group (for some prime p) and if $|B| > p$ then H may be expressed as a central product as in Proposition 9.

Clearly all the hypotheses of the theorem are satisfied when $A_1 = N_A(K_1)$ replaces A and K_1 replaces H. Since $K_1 \neq H$ we may apply induction to conclude that if χ_1 is any irreducible constituent of the restriction of χ to K_1 then χ_1 extends to A_1K_1. Let V_1 be an irreducible $\mathbb{C}A_1K_1$-module affording this character.

Choose representatives $l_1, l_2, \ldots, l_m$ for the left cosets l_iA_1 of A_1 in A, so that $K_i^{l_i} = K_1$ $(i = 1, 2, \ldots, m)$. Let V_i be a vector space over $\mathbb{C}$ isomorphic to V_1 and let $\rho_i : V_i \to V_1$ be an isomorphism. Then V_i is an irreducible $\mathbb{C}K_i$-module under

$$xv = \rho_i^{-1}x^{l_i}(\rho_iv) \quad (x \in K_i, v \in V_i) .$$

Clearly, the tensor product $V_1 \otimes V_2 \otimes \ldots \otimes V_m$ is an irreducible $\mathbb{C}H$-module under

$$x_1x_2 \ldots x_m(v_1 \otimes v_2 \otimes \ldots \otimes v_m) = x_1v_1 \otimes x_2v_2 \otimes \ldots \otimes x_mv_m \quad (x_i \in K_i, v_i \in V_i),$$

and this module affords the character χ (since χ is the unique irreducible constituent of χ_1^H). But now $V_1 \otimes V_2 \otimes \ldots \otimes V_m$ becomes a $\mathbb{C}AH$-module if we define for $a \in A$,

$$a(v_1 \otimes v_2 \otimes \ldots \otimes v_m) = u_1 \otimes u_2 \otimes \ldots \otimes u_m$$

where u_i is defined by

$$\rho_i u_i = l_i^{-1} a l_j (\rho_j v_j) \quad (i = 1, 2, \ldots, m),$$

j being the unique index such that $al_jA_1 = l_iA_1$. Thus we have contradicted the assumption that χ does not extend, and the proposition is proved. (The above construction appears in [1].)

We now complete the proof of the theorem by deriving a final contradiction. Let $L_1 = C_L(B)$. Then $[L_1, H]Z(H) \neq H$ and is A-invariant; so by Proposition 6, $[L_1, H] \leq Z(H)$. Now

$$[L_1, H, B] = [B, L_1, H] = 1$$

hence

$$1 = [H, B, L_1] = [H, L_1]$$

by hypothesis (3) of Theorem 1. But $L_1 \leq \mathrm{aut}(H)$; so $L_1 = 1$. Since $|B| = p$ and $C_L(B) = 1$ it follows that L, being a group of automorphisms of B, is cyclic. By Lemma 2 *(a)*, χ extends to LH, and using Lemma 2 *(b)* and *(c)* as in the proof of Proposition 7, χ extends to AH. This is the required contradiction.

References

[1] Everett C. Dade, "Characters of groups with normal extra special subgroups", *Math. Z.* 152 (1976), 1-31.

[2] P.X. Gallagher, "Group characters and normal Hall subgroups", *Nagoya Math. J.* 21 (1962), 223-230.

[3] George Glauberman, "Correspondence of characters for relatively prime operator groups", *Canad. J. Math.* 20 (1968), 1465-1488.

[4] Daniel Gorenstein, *Finite Groups* (Harper and Row, New York, Evanston, London, 1968).

[5] I.M. Isaacs, "Invariant and extendible group characters", *Illinois J. Math.* **14** (1970), 70-75.

[6] Thomas R. Wolf, "Characters of p'-degree in solvable groups", *Pacific J. Math.* **74** (1978), 267-271.

Department of Pure Mathematics,
University of Sydney,
Sydney,
New South Wales.

PROC. 18th SRI, 16-02
CANBERRA 1978, 8-46 (16A38)

SOME RECENT DEVELOPMENTS IN THE THEORY OF ALGEBRAS WITH POLYNOMIAL IDENTITIES

N. Jacobson

The author gave six lectures at the Summer Research Institute: two survey lectures and four lectures on algebras with polynomial identities. The survey lectures had the titles: *Development of the Concept of a Jordan Structure* and *History of Algebras with Polynomial Identity*. The author decided not to include these lectures in the Proceedings since these would have overlapped substantially with surveys that have appeared or are about to appear, notably, a survey article on Jordan algebras by McCrimmon that will appear shortly in the Bulletin of the American Mathematical Society, a survey lecture "*PI*-algebras" by the author appearing in the Proceedings of a ring theory conference at University of Oklahoma published in 1973 by M. Dekker, and a survey article "Polynomial identities" by S.A. Amitsur in Israel Journal of Mathematics, 19 (1974), 183-199.

The lectures on recent developments in the *PI*-theory have been written up with the following titles:

I Razmyslov's central polynomial,

II The Artin-Procesi Theorem,

III On Shirshov's local finiteness theorems.

Central polynomials for the complete matrix algebra $M_n(K)$, K a commutative ring, have played an important role in recent developments of the *PI*-theory. For example, they can be used to give an improved formulation and proof of the main structure theorem for prime *PI*-algebras. The first construction of central polynomials for

PI-algebras was given in 1972 by Formanek. In Chapter I we present an alternative construction due to Razmyslov that has the advantage that the central polynomial is multilinear and is alternating in n^2 of the arguments. Razmyslov's central polynomial plays an important role in II that is concerned with a theorem connecting *PI*-algebras with Azumaya algebras.

In III we give an account (with some improvements) of the best positive results that have appeared to date on a problem in ring theory, Kurosch's problem, that is analogous to Burnside's problem in group theory.

CHAPTER I

RAZMYSLOV'S CENTRAL POLYNOMIAL

Let A be an algebra (associative with unit) over a commutative ring K and let $K\{x_1, x_2, \ldots\}$ be the free associative algebra generated by $x_1, x_2, \ldots$. An element $f(x_1, \ldots, x_m) \in K\{x_1, x_2, \ldots\}$ is called a *central polynomial* for A if $f(a_1, \ldots, a_m) \in C = C(A)$ the center of A for all $a_i \in A$ and there exist $a_1, \ldots, a_m$ such that $f(a_1, \ldots, a_m) \neq 0$. In other words,

$$[f(x_1, \ldots, x_m), x_{m+1}]$$

is an identity for A but $f(x_1, \ldots, x_m)$ is not. We shall give a construction due to Razmyslov, of a multilinear central polynomial for $M_n(K)$, the algebra of $n \times n$ matrices over K .

We require first some elementary results on $M_n(K)$.

LEMMA 1. *The trace bilinear form* $t(A, B) = \operatorname{tr} AB$ *is non-degenerate on* $M_n(K)$: $t(A, B) = 0$ *for all* B *implies* $A = 0$.

Proof. If $A = (a_{ij})$ and $\{e_{kl} \mid 1 \le k, l \le n\}$ are the usual matrix units for $M_n(K)$, then $t(A, e_{kl}) = a_{lk}$. Hence $t(A, e_{kl}) = 0$ for all k, l implies $A = 0$. □

$M_n(K)$ is a free module with base of n^2 elements. Hence any K-endomorphism of $M_n(K)$ determines an $n^2 \times n^2$ matrix relative to a base. The trace of this matrix is independent of the choice of base and is called the *trace of the endomorphism*. In particular, we consider the endomorphism $U \mapsto AUB$ defined by a given pair of matrices A and B. Then we have

LEMMA 2. *The trace of the endomorphism* $U \mapsto AUB$ *of* $M_n(K)$ *is* $(\mathrm{tr}\, A)(\mathrm{tr}\, B)$.

This can be verified by direct calculation using the base $\{e_{ij}\}$ of matrix units. We omit this simple calculation.

We note next that if A is any algebra and $\mathsf{A}^e = \mathsf{A} \otimes_K \mathsf{A}^o$ where A^o is the opposite algebra then we have the involution $\sum a_i \otimes b_i \mapsto \sum b_i \otimes a_i$ in A^e. Moreover, A is an A^e-module relative to the action $\left(\sum a_i \otimes b_i\right)x = \sum a_i x b_i$. This action is by K-endomorphisms. For a certain class of algebra, the Azumaya algebras, the representation defined by this action is faithful and the image is $\mathrm{End}_K \mathsf{A}$. We can verify this for $\mathsf{A} = M_n(K)$:

LEMMA 3. *The map sending* $\sum a_i \otimes b_i$ *into the endomorphism* $x \mapsto \sum a_i x b_i$ *of* $M_n(K)$ *into itself is an isomorphism onto* $\mathrm{End}_K M_n(K)$.

Proof. Since both $M_n(K) \otimes M_n(K)^o$ and $\mathrm{End}_K M_n(K)$ are free K-modules of rank n^4 it suffices to show that the homomorphism is surjective. For this it is enough to show that for any pair $(i, j), (k, l)$ there exists a map of the form $U \mapsto \sum A_i U B_i$ sending e_{ij} into e_{kl} and $e_{i'j'}$ into 0 for every $(i', j') \neq (i, j)$. It is clear that $U \mapsto e_{ki} U e_{jl}$ does this. □

In the situation in which we have an isomorphism of A^e onto $\mathrm{End}_K \mathsf{A}$ defined as above we can transfer the involution $\sum a_i \otimes b_i \mapsto \sum b_i \otimes a_i$ to an involution in $\mathrm{End}_K \mathsf{A}$. We denote this involution as $l \mapsto l^*$.

Now let $X = K[\xi_{11}, \ldots, \xi_{ij}, \ldots, \xi_{nn}]$ where the ξ_{ij} are indeterminates and consider $M_n(X)$ which is the same thing as $X \otimes_K M_n(K)$. We can regard $M_n(X)$ as $M_n(K)^e$-module obtained by restricting the action from $M_n(X)^e$ to the subalgebra $M_n(K)^e$. Consider the "generic" matrix $X = (\xi_{ij}) \in M_n(X)$ and the cyclic $M_n(K)^e$-module generated by X . This is a free K-module with base $\{\xi_{ij}e_{kl} \mid i, j, k, l = 1, \ldots, n\}$. If $l_1, l_2 \in M_n(K)^e$ and $l_1(X) = l_2(X)$ then $l_1(e_{ij}) = l_2(e_{ij})$ for all i, j and hence $l_1 = l_2$. It follows that we have a well defined K-endomorphism A_X of $M_n(K)^eX$ such that

(1) $$A_X : l(X) \mapsto l^*(X) .$$

We call this the *Razmyslov transposition* in $M_n(K)^eX$. Since

$$\xi_{ij}e_{kl} = e_{ki}Xe_{jl}$$

we have

(2) $$A_X\xi_{ij}e_{kl} = e_{jl}Xe_{ki} = \xi_{lk}e_{ji} .$$

Hence the Razmyslov transposition can also be defined directly as the map

(3) $$\sum a_{ijkl}\xi_{ij}e_{kl} \mapsto \sum a_{ijkl}\xi_{lk}e_{ji}$$

in $M_n(K)^eX$.

LEMMA 4. *If* $H \in M_n(K)$ *then* $(\mathrm{tr}\, X)H \in M_n(K)^eX$ *and*

(4) $$A_X(\mathrm{tr}\, X)H = \mathrm{tr}(XH)1 .$$

Proof. Since $\mathrm{tr}\, X = \sum \xi_{ii}$, $(\mathrm{tr}\, X)H$ is a K-linear combination of the elements $\xi_{ii}e_{kl}$ and so is contained in $M_n(K)^eX$. Write $H = \sum h_{jk}e_{jk}$, $h_{jk} \in K$. Then $(\mathrm{tr}\, X)H = \sum_{i,j,k} h_{jk}\xi_{ii}e_{jk}$. Hence, by (2),

$$\begin{aligned} A_X(\mathrm{tr}\, X)H &= \sum_{i,j,k} h_{jk}\xi_{kj}e_{ii} = (\mathrm{tr}\, HX)1 \\ &= \mathrm{tr}(XH)1 . \quad \square \end{aligned}$$

We now consider the free algebra $K\{x_1, x_2, \ldots; y_1, y_2, \ldots\}$ generated by $x_1, x_2, \ldots; y_1, y_2, \ldots$. Let $h(x_1, \ldots, x_r, y_1, \ldots, y_s)$ be multilinear and alternating in the x's . Let $A = (a_{ij}) \in M_r(K)$ and $x_i' = \sum a_{ij}x_j$. Then direct substitution shows that

(5) $$h(x_1', \ldots, x_r', y_1, \ldots, y_s) = \det(A)h(x_1, \ldots, x_r, y_1, \ldots, y_s) .$$

Let λ be an indeterminate and replace K by $K[\lambda]$, A by $\lambda 1 - A$. Then comparing coefficients of the powers of λ in the relation obtained from (5) we obtain a number of relations like (5). In particular, taking the coefficient of λ^{n-1} we obtain

$$(6)\quad h(x_1', x_2, \ldots, x_r, y_1, \ldots, y_s) + h(x_1, x_2', x_3, \ldots, x_r, y_1, \ldots, y_s) + \ldots + h(x_1, \ldots, x_{r-1}, x_r', y_1, \ldots, y_s) = \operatorname{tr}(A)h(x_1, \ldots, x_r, y_1, \ldots, y_s) .$$

This holds in particular for the *Capelli polynomial*

$$(7)\quad h(x_1, \ldots, x_{n^2}, y_1, \ldots, y_{n^2-1}) = \sum_{\pi\in\Sigma_{n^2}} (\operatorname{sg} \pi)x_{\pi 1}y_1x_{\pi 2}y_2 \ldots x_{\pi n^2}$$

of degree $2n^2 - 1$. For this we have

LEMMA 5. *If* $(A_1, \ldots, A_{n^2})$ *is a base for* $M_n(K)$ *then we can choose* $(B_1, \ldots, B_{n^2-1})$ *so that*

$$(8)\quad h(A_1, \ldots, A_{n^2}, B_1, \ldots, B_{n^2-1}) \neq 0 .$$

Proof. Using (5) it suffices to prove the result for the base $\{e_{ij}\}$. It is clear that for any ordering of this base, say, as $A_1 = e_{ij}, \ldots, A_{n^2} = e_{kl}$ there is only one choice of $B_1, \ldots, B_{n^2-1}$ in $\{e_{ij}\}$ so that $A_1B_1A_2B_2 \ldots B_{n^2-1}A_{n^2} \neq 0$. Then the product is e_{il} and $h(A_1, A_2, \ldots, A_{n^2}, B_1, \ldots, B_{n^2-1}) = e_{il} \neq 0$. □

LEMMA 6. *Let* h *be the Capelli polynomial of degree* $2n^2 - 1$ *and let* $A_1, \ldots, A_{n^2}, B_1, \ldots, B_{n^2-1}, C, D \in M_n(K)$. *Then*

$$(9)\quad h(CA_1D, A_2, \ldots, A_{n^2}, B_1, \ldots, B_{n^2-1}) + h(A_1, CA_2D, A_3, \ldots, A_{n^2}, B_1, \ldots, B_{n^2-1}) + \ldots + h(A_1, \ldots, A_{n^2-1}, CA_{n^2}D, B_1, \ldots, B_{n^2-1}) = (\operatorname{tr} C)(\operatorname{tr} D)h(A_1, \ldots, A_{n^2}, B_1, \ldots, B_{n^2-1}) .$$

Proof. Since the relation to be proved is multilinear in all of the variables $A_1, \ldots, A_{n^2}, B_1, \ldots, B_{n^2-1}, C, D$ and since $M_n(K) \cong K \otimes_{\mathbb{Z}} M_n(\mathbb{Z})$, it suffices to prove (9) for $M_n(\mathbb{Z})$. Now (9) will follow for $M_n(\mathbb{Z})$ if we can prove it for $M_n(\mathbb{Q})$. Hence it suffices to take $K = \mathbb{Q}$. Using the Zariski topology on $M_n(\mathbb{Q})$ it suffices to prove (9) on the Zariski open subset on which $(A_1, \ldots, A_{n^2})$ is a base. (This is defined by the non-vanishing of the determinant of the $n^2 \times n^2$ matrix expressing the A_i in terms of the base $\{e_{ij}\}$.) Consider the linear transformation $U \mapsto CUD$ of $M_n(\mathbb{Q})$. We obtain the matrix of this linear transformation by writing

(10) $$CA_iD = \sum_{j=1}^{n^2} p_{ij}A_j \ , \quad 1 \le i \le n^2 .$$

By (6), the left hand side of (9) equals $\mathrm{tr}(p)h(A_1, \ldots, A_{n^2}, B_1, \ldots, B_{n^2-1})$ where $(p) = (p_{ij})$. By Lemma 2, $\mathrm{tr}(p) = (\mathrm{tr}\ C)(\mathrm{tr}\ D)$. Hence (9) holds. □

We can now define Razmyslov's central polynomial.

THEOREM 1 (Razmyslov). *Let h be the Capelli polynomial of degree $2n^2 - 1$ and let H be the polynomial obtained from*

(11) $$h(y_{n^2}x_1y_{n^2+1}, x_2, \ldots, x_{n^2}, y_1, \ldots, y_{n^2-1}) + h(x_1, y_{n^2}x_2y_{n^2+1}, \ldots, x_{n^2}, y_1, \ldots, y_{n^2-1}) + \ldots + h(x_1, \ldots, x_{n^2-1}, y_{n^2}x_{n^2}y_{n^2+1}, y_1, \ldots, y_{n^2-1})$$

by "pivoting on y_{n^2}": replacing every monomial $wy_{n^2}z$ occurring in (11) *by $zy_{n^2}w$. Then for any $A_1, \ldots, A_{n^2}, B_1, \ldots, B_{n^2+1} \in M_n(K)$ we have the relation*

(12) $$H(A_1, \ldots, A_{n^2}, B_1, \ldots, B_{n^2+1}) = \mathrm{tr}\ B_{n^2+1}\ \mathrm{tr}(B_{n^2}h(A_1, \ldots, A_{n^2}, B_1, \ldots, B_{n^2-1}))1 .$$

Moreover, H is a central polynomial for $M_n(K)$.

Proof. We consider $M_n(X)$ where $X = K[\xi_{11}, \ldots, \xi_{ij}, \ldots, \xi_{nn}]$ and let $X = (\xi_{ij})$. By (9), we have

(13) $$h(XA_1B_{n^2+1}, A_2, \ldots, A_{n^2}, B_1, \ldots, B_{n^2-1}) + \ldots + h(A_1, \ldots, A_{n^2-1}, XA_{n^2}B_{n^2+1}, B_1, \ldots, B_{n^2-1}) = (\mathrm{tr}\ B_{n^2+1})(\mathrm{tr}\ X)h(A_1, \ldots, A_{n^2}, B_1, \ldots, B_{n^2-1}) .$$

Next we apply the Razmyslov transposition A_X to both sides of this equation. By (4) we obtain

$$H(A_1, \ldots, A_{n^2}, B_1, \ldots, B_{n^2-1}, X, B_{n^2+1}) = \mathrm{tr}\ B_{n^2+1}\ \mathrm{tr}(Xh(A_1, \ldots, A_{n^2}, B_1, \ldots, B_{n^2-1}))1 .$$

Specializing $X \mapsto B_{n^2}$ gives (12). It is clear from (12) that $H(A_1, \ldots, A_{n^2}, B_1, \ldots, B_{n^2+1}) \in K1$ for all choices of the A_i, B_j . It remains to show that the A_i, B_j can be chosen so that $H(A_1, \ldots, A_{n^2}, B_1, \ldots, B_{n^2+1}) \ne 0$. To do this we choose B_{n^2+1} so that $\mathrm{tr}\ B_{n^2+1} = 1$, $(A_1, \ldots, A_{n^2})$ to be a base for $M_n(K)$. Then, by Lemma 5, we can choose $B_1, \ldots, B_{n^2-1}$ so that

$h(A_1, \ldots, A_{n^2}, B_1, \ldots, B_{n^2-1}) \neq 0$. By Lemma 1, we can choose B_{n^2} so that $\operatorname{tr} B_{n^2}h(A_1, \ldots, A_{n^2}, B_1, \ldots, B_{n^2-1}) \neq 0$. These choices give a non-zero value for H . □

We shall now extend Theorem 1 to prime algebras satisfying proper identities, where a polynomial $f(x_1, \ldots, x_m)$ is called *proper* for the algebra A if for some coefficient α in $f(x_1, \ldots, x_m)$ we have $\alpha A \neq 0$ or, equivalently, $\alpha 1 \neq 0$, for the unit 1 of A . We define the *PI-degree* of A to be the least integer d for which there exists a proper identity of degree d . If A is a prime algebra satisfying a proper identity and C is the center of A then C is a domain and, by the structure theorem for prime *PI*-algebras, if F is the field of fractions of C then $A \hookrightarrow A_F = F \otimes_C A$ and A_F is finite dimensional central simple over F . Moreover, A and A_F have the same identities and $[A_F : F] = n^2$ (Jacobson [1], p. 57). If $\overline{F}$ is the algebraic closure of F then $A_{\overline{F}} = \overline{F} \otimes_C A \cong \overline{F} \otimes_F A_F \cong M_n(\overline{F})$. By the Amitsur-Levitzki Theorem S_{2n} is an identity for $A_{\overline{F}}$ and since $A \hookrightarrow A_{\overline{F}}$, S_{2n} is an identity for A . On the other hand, if f is a proper identity of degree less than $2n$ for A then A satisfies a proper multilinear identity of degree less than $2n$. Then $M_n(\overline{F}) \cong A_{\overline{F}}$ satifies this identity. This is impossible (Jacobson [1], p. 16). Hence, we see that the *PI*-degree of A is $2n$ if $[A_F : F] = n^2$.

We now prove

THEOREM 2. *If* A *is a prime algebra of PI-degree* $2n$ *and* C *is the center of* A *then* A *and* $M_n(C)$ *have the same identities and the same central polynomials.*

Proof. Since $f(x_1, \ldots, x_m)$ is a central polynomial if and only if it is not an identity but $[f(x_1, \ldots, x_m), x_{m+1}]$ is an identity the second statement follows from the first. Since A_F and A have the same identities and proper identities and $M_n(C)$ and $M_n(F)$ have the same identities and proper identities it suffices to prove that this is the case also for A_F and $M_n(F)$. If $\alpha \in K$ then $\alpha 1 \in C \subset F$. Hence any proper identity for A yields a non-zero identity with coefficients in F for A_F . Hence it suffices to assume $K = F$ in considering A_F and $M_n(F)$. We distinguish two cases:

I F finite and

II F infinite.

In the first case, by Wedderburn's Theorem on finite division rings, $A_F = M_n(F)$. Hence the result holds in this case. Now assume F infinite. We prove first that if

A is a finite dimensional algebra over an infinite field F and E is an extension field of F then A and $A_E = E \otimes_F A$ have the same identities over F. Let $(a_1, \ldots, a_r)$ be a base for A/F, hence for A_E/E. Let $f(x_1, \ldots, x_m) \in F\{x_1, x_2, \ldots\}$ and let $\xi_j^{(i)}$, $1 \le i \le m$, $1 \le j \le r$, be indeterminates. Put $X = F\left[\xi_1^{(1)}, \ldots, \xi_r^{(1)}, \ldots, \xi_r^{(m)}\right]$ and consider the algebra A_X and the "generic" elements $x^{(i)} = \sum_j \xi_j^{(i)} a_j$ of this algebra. Since $(a_1, \ldots, a_r)$ is a base for A_X/X we can write $f(x^{(1)}, \ldots, x^{(m)}) = \sum \varphi_l\left(\xi_1^{(1)}, \ldots, \xi_r^{(m)}\right) a_l$. Then f is an identity for A if and only if $\varphi_l\left(\alpha_1^{(1)}, \ldots, \alpha_r^{(m)}\right) = 0$ for all $\alpha_j^{(i)} \in F$ and all $l = 1, 2, \ldots, m$. Since F is infinite this occurs if and only if every $\varphi_l = 0$. Since $(a_1, \ldots, a_r)$ is a base for A_E/E the same conditions obtain for A_E. Hence A and A_E have the same F-identities. We apply this to $A_{\bar{F}}$, A as before, taking $E = \bar{F}$. This shows that $A_{\bar{F}}$ and $M_n(\bar{F})$ have the same F-identities. Similarly $M_n(F)$ and $M_n(\bar{F}) \cong M_n(F)_{\bar{F}}$ have the same F-identities. Hence this holds also for A_F and $M_n(F)$. □

An immediate consequence of this result and Theorem 1 is the following.

COROLLARY. *If* A *is a prime algebra of PI-degree* $2n$ *then the Razmyslov polynomial of degree* $2n^2 + 1$ *is a central polynomial for* A.

References

Nathan Jacobson

[1] *PI-Algebras. An Introduction* (Lecture Notes in Mathematics, 441. Springer-Verlag, Berlin, Heidelberg, New York, 1975).

Ju.P. Razmyslov

[1] "Trace identities of full matrix algebras over a field of characteristic zero", *Math. USSR Izv.* 8 (1974), no. 4., 727-760.

CHAPTER II

THE ARTIN-PROCESI THEOREM

In 1969, Michael Artin [1] established a surprising connection between two important areas of ring theory: Azumaya algebras and algebras with polynomial identity. He gave a characterization by identities of the most interesting class of Azumaya algebras, the Azumaya algebras that have a rank.

There are a number of equivalent definitions of an Azumaya algebra that can be given. The usual one is: An algebra A over a commutative ring K is called an *Azumaya algebra* if A regarded in the natural way as $A^e = A \otimes_K A^o$ module is projective and the map $\alpha \mapsto \alpha 1$ of K into A is an isomorphism onto the center $C(A)$. If the first condition alone holds then the algebra is called *separable*. The second condition states that $K1$ is the center of A and $\alpha 1 = 0$ for $\alpha \in K$ implies $\alpha = 0$. If this condition holds then A is called *central*. Thus central plus separable equals Azumaya.

If K is a field it can be shown that A is separable over K in the above sense if and only if A is classically separable, that is, A is finite dimensional over K and $A_{\overline{K}} = \overline{K} \otimes_K A$ is semi-simple for $\overline{K}$ the algebraic closure of K. Moreover, A is Azumaya over a field K if and only if A is finite dimensional central simple over K. Thus the theory of separable algebras over commutative rings and the theory of Azumaya algebras constitute natural generalizations of the classical theories of finite dimensional separable and central simple algebras.

We need to recall the definition of rank of a finitely generated projective module M over a commutative ring K. Let P be a prime ideal in K and let K_P be the localization of K at P. This is a local ring, that is, it has a unique maximal ideal P_P. Moreover, we can form the localization M_P of M at P, that can be defined as $K_P \otimes_K M$. This is finitely generated projective over K_P and since K_P is local it follows that M_P is a free module of finite rank n_P over K_P. We say that M *has a rank over* K if n_P is constant as P ranges over the set of prime ideals of K. In this case $n = n_P$ is called *the rank* of the finitely generated projective module M over K.

It can be shown that the rank is defined if K contains no idempotents not equal to $0, 1$. More generally, it can be shown that if M is finitely generated projective over K then there are only a finite number of distinct ranks n_P for P a prime ideal in K and if these ranks are $n_1, \ldots, n_s$ then $K = K_1 \oplus \ldots \oplus K_s$ where the K_i are ideals and $M = K_1M \oplus \ldots \oplus K_sM$ where K_iM has rank n_i over K_i. It can be shown also that if M has a rank n over K and K' is a commutative algebra over K then $M_{K'} = K' \otimes_K M$ has rank n over K'.

It can be shown that if A is an Azumaya algebra over K then A is finitely generated projective over K. Moreover, if K' is a commutative algebra over K then $A_{K'}$ is Azumaya. Let P be a prime ideal of K. Then $A_P = K_P \otimes_K A$ is Azumaya over K_P and its rank over K_P is the same as that of $(K_P/P_P) \otimes_{K_P} A_P$ over K_P/P_P. Since K_P/P_P is a field and $(K_P/P_P) \otimes_{K_P} A_P$ is Azumaya over K_P/P_P its rank is the dimensionality and this is a square. It follows that the rank n_P of A_P over K_P is a square. Hence if A has a rank over K then this is a square. It can be shown also that if A is a ring that is Azumaya over its center C then A has a unique decomposition as direct sum of a finite number of rings A_i that are Azumaya over their centers and have ranks over their centers (Bix [1]).

The theorem of Artin-Procesi that we shall prove characterizes Azumaya algebras that have a rank.

DEFINITION 1. A ring A is called an A_n*-ring* if A (regarded as an algebra over $\mathbb{Z}$) satisfies every identity of $M_n(\mathbb{Z})$ but no non-zero homomorphic image of A satisfies every identity of $M_{n-1}(\mathbb{Z})$.

Then we have the

ARTIN-PROCESI THEOREM. *A ring* A *is an* A_n*-ring if and only if it is an Azumaya algebra of rank* n^2 *over its center.*

This was proved by Artin under the hypothesis that A is an algebra over a field. This restriction was removed by Procesi. After central polynomials became available simpler proofs of the result were given by Amitsur, by Rowen, by Goldie and finally by Schelter. We shall give Schelter's proof which, it should be noted, was confined to proving A_n implies Azumaya for prime rings. While the general case can be reduced to that of prime rings, it is simpler to avoid this reduction and prove directly that A_n implies Azumaya for arbitrary A. We shall do this. The implication that if A is Azumaya of rank n^2 then A is an A_n-ring can be deduced from a deep theorem on splitting rings of Azumaya algebras (see Knus, Ojanguren, p. 104). We shall give a more elementary proof via a series of reductions. This may be of interest to readers who are not experts in the theory of Azumaya algebras.

Following Schelter, we base the proof that A_n implies Azumaya on the following well known characterization of Azumaya algebras: An algebra A over K is Azumaya if and only if A is finitely generated projective and faithful over K and the canonical homomorphism of A^e into $\mathrm{End}_K A$ is an isomorphism. The statement that A is faithful over K means that if $\alpha \in K$ and $\alpha A = 0$ then $\alpha = 0$. A characterization of projective modules that we shall use is in terms of a dual basis (or "projective coordinates"): A module M over a ring R is projective if and only if there exists a set $\{(x_i, f_i)\}$ of pairs (x_i, f_i) where $x_i \in M$, $f_i \in M^* = \mathrm{Hom}_R(M, R)$ such that for any $x \in M$, $f_i(x) = 0$ for all but a finite number of i and $x = \sum f_i(x)x_i$. The set $\{(x_i, f_i)\}$ is called a *dual basis* for M. M is finitely generated projective if and only if it has a finite dual basis.

We shall need the following

LEMMA 1. *Let M be finitely generated projective over a commutative ring K and let $\{(x_i, f_i) \mid 1 \le i \le n\}$ be a dual basis for M. Let E_{ij} denote the endomorphism $x \mapsto f_i(x)x_j$. Then the n^2 endomorphisms E_{ij} span $\mathrm{End}_K M$ over K.*

Proof. Let $L \in \mathrm{End}_K M$ and suppose $Lx_i = \sum_j \lambda_{ij}x_j$, $\lambda_{ij} \in K$. Then

$$Lx = L\left(\sum f_i(x)x_i\right) = \sum \lambda_{ij}f_i(x)x_j$$
$$= \sum \lambda_{ij}E_{ij}x .$$

Hence $L = \sum \lambda_{ij}E_{ij}$.

We shall require also

LEMMA 2. *There exists an element $H(x_1, \ldots, x_{n^2}, y_1, \ldots, y_s)$ in the free associative algebra $\mathbb{Z}\{x_1, x_2, \ldots; y_1, y_2, \ldots\}$ such that:*

(1) *H is multilinear in the* x*'s and the* y*'s ;*

(2) *H is alternating in the* x*'s; and*

(3) *H is central for* $M_n(K)$ *for any commutative ring* K .

This is a consequence of Razmyslov's theorem that was proved in the preceding chapter.

Let H be as in Lemma 2 and put

$$H'(x_1, \ldots, x_{n^2+1}, y_1, \ldots, y_s) = H(x_1, \ldots, x_{n^2}, y_1, \ldots, y_s)x_{n^2+1} + \sum_{j=1}^{n^2} (-1)^{n^2+j+1} H(x_1, \ldots, \hat{x}_j, \ldots, x_{n^2+1}, y_1, \ldots, y_s)x_j .$$

It is readily verified that H' is alternating and multilinear in $x_1, \ldots, x_{n^2+1}$. Hence H' is an identity for $M_n(\mathbb{Z})$.

We shall now give the

Proof of the Artin-Procesi Theorem. We begin with the proof of

$$A_n \Rightarrow \textit{Azumaya of rank } n^2 \textit{ over its center.}$$

Let A be a ring satisfying A_n and let $\bar{A}$ be a simple homomorphic image of A . Then $\bar{A}$ satisfies the standard identity S_{2n} and hence, by Kaplansky's theorem, $[\bar{A} : C(\bar{A})] = m^2 \leq n^2$. We claim that the polynomial H is not an identity for $\bar{A}$. Otherwise, if $\bar{F}$ is the algebraic closure of $C(\bar{A})$ then H is an identity for $M_n(\bar{F}) \cong \bar{F} \otimes_{C(\bar{A})} \bar{A}$. Since H is central for $M_n(\bar{F})$ this implies that $m < n$. Now any $\mathbb{Z}$-identity for $M_m(\mathbb{Z})$ is an identity for $M_m(\mathbb{Q})$ and since $\mathbb{Q}$ is infinite it follows, as in the proof of Theorem 2 of the last chapter, that any $\mathbb{Z}$-identity for $M_m(\mathbb{Z})$ is an identity for $M_m(\bar{F})$ and hence for $\bar{A}$. Thus every identity for $M_{n-1}(\mathbb{Z})$ is an identity for $\bar{A}$ contrary to the hypothesis. Hence H is not an identity for $\bar{A}$. On the other hand, $[H, x_{n^2+1}]$ is an identity for $M_n(\mathbb{Z})$, hence for A , so H is a central polynomial for A .

Now let B be the ideal in A generated by all the values $H(a_1, \ldots, a_{n^2}, b_1, \ldots, b_s)$, $a_i, b_j \in A$. We claim $B = A$. Otherwise, B is contained in a maximal ideal M of A and $\bar{A} = A/M$ is a simple homomorphic image of A . It is clear that H is an identity for $\bar{A}$ contrary to what we showed before. Thus $B = A$. Since H is central, B is the set of elements of the form

(2) $$\sum_k d_k H\left[a_1^{(k)}, \ldots, a_{n^2}^{(k)}, b_1^{(k)}, \ldots, b_s^{(k)}\right]$$

where the a's, b's and d's are in A. Since B = A we have $a_i^{(k)}, b_i^{(k)}, d_k$ such that

(3) $$\sum_k d_k H\left[a_1^{(k)}, \ldots, a_{n^2}^{(k)}, b_1^{(k)}, \ldots, b_s^{(k)}\right] = 1 .$$

Hence for any $a \in A$ we have

(4) $$a = \sum H\left[a_1^{(k)}, \ldots, a_{n^2}^{(k)}, b_1^{(k)}, \ldots, b_s^{(k)}\right] a d_k .$$

Now put

(5) $$f_{kj}(a) = (-1)^{j+n^2} H\left[a_1^{(k)}, \ldots, \widehat{a_j^{(k)}}, \ldots, a_{n^2}^{(k)}, a, b_1^{(k)}, \ldots, b_s^{(k)}\right] .$$

Since H' is an identity for $M_n(\mathbb{Z})$, hence for A, by the definition of H' we have

(6) $$H\left[a_1^{(k)}, \ldots, a_{n^2}^{(k)}, b_1^{(k)}, \ldots, b_s^{(k)}\right] a = \sum_j f_{kj}(a) a_j^{(k)} .$$

Hence, by (4), we have

(7) $$a = \sum_{k,j} f_{kj}(a) a_j^{(k)} d_k .$$

Since f_{kj} is a $C(A)$-linear map of A into $C(A)$ it follows that $\left\{\left[a_j^{(k)} d_k, f_{kj}\right]\right\}$ is a dual base for A as $C = C(A)$-module. Hence A is finitely generated projective over C. Evidently A is C-faithful also.

We show next that the canonical map φ of A^e into $\mathrm{End}_K A$ is surjective. By Lemma 1, it suffices to show that for any k, j and any b, the map $a \mapsto f_{kj}(a)b$ has the form $a \mapsto \sum e_i a e_i'$ for suitable e_i, e_i' in A. This is clear from the definition of the f_{kj} given in (5).

Next we have to show that φ is injective. Suppose $\varphi\left[\sum e_i \otimes e_i'\right] = 0$, so $\sum_i e_i a e_i' = 0$ for all $a \in A$. Write $f_{kj} = \varphi\left[\sum_l u_{kjl} \otimes u_{kjl}'\right]$. Then, by (7), we have

$$a = \sum_{k,j,l} u_{kjl} a u_{kjl}' a_j^{(k)} d_k .$$

Hence

$$\begin{aligned}\sum e_i \otimes e_i' &= \sum_{i,j,k,l} u_{kjl} e_i u_{kjl}' a_j^{(k)} d_k \otimes e_i' \\ &= \sum a_j^{(k)} d_k \otimes u_{kjl} e_i u_{kjl}' e_i' \\ &= \sum_{k,j,l} a_j^{(k)} d_k \otimes u_{kjl} \sum_i e_i u_{kjl}' e_i' \\ &= 0\end{aligned}$$

since $\sum_i e_i u_{kjl}' e_i' = 0$. Thus φ is injective and so φ is an isomorphism. Then A is Azumaya by the conditions we noted before.

We show next that A has rank n^2 over C . Let P be a prime ideal of C . We have seen that the rank n_P of $A_P = C_P \otimes_C A$ over C_P is the same as the rank of $(C_P/P_P) \otimes_{C_P} A_P$ over the field C_P/P_P , and this is a square m^2 . Now A satisfies the standard identity S_{2n} . Hence S_{2n} is an identity for A_P and for $(C_P/P_P) \otimes_{C_P} A_P$. Suppose $m < n$. Then it follows from Theorem 2 of the last chapter that $(C_P/P_P) \otimes_{C_P} A_P$ satisfies every identity of $M_m(C_P/P_P)$ and hence every identity of $M_m(\mathbb{Z})$. Then $(C_P/P_P) \otimes_{C_P} A_P$ satisfies every identity of $M_{n-1}(\mathbb{Z})$. On the other hand $(C_P/P_P) \otimes_{C_P} A_P = (C_P/P_P) \otimes_{C_P} (C_P \otimes_C A) \cong (C_P/P_P) \otimes_C A$ and we have the canonical homomorphism $a \mapsto 1 \otimes a$ of A into $(C_P/P_P) \otimes_C A$. This gives a non-zero homomorphic image $\bar{A}$ of A that satisfies every identity of $M_{n-1}(\mathbb{Z})$ contrary to hypothesis. Hence $n_P = n^2$. Since this holds for every prime ideal P of C , A has rank n^2 over C . □

We prove next the converse:

Azumaya of rank n^2 *over the center* $\Rightarrow A_n$.

The second condition in the definition of A_n-ring is easily established for any Azumaya algebra A of rank n^2 . For it is known that we have a bijection $I \mapsto IA$ of the set of ideals of the center C of A with the set of ideals of A and $A/IA \cong (C/I) \otimes_C A$. Hence any non-zero homomorphic image of A is Azumaya of rank n^2 . Now suppose $\bar{A}$ is a non-zero homomorphic image that satisfies every identity of $M_{n-1}(\mathbb{Z})$. Then we may assume $\bar{A}$ is simple. Then $\bar{A}$ is n^2

dimensional over its center and hence $\bar{A}$ does not satisfy $S_{2(n-1)}$. This contradiction proves the result.

It remains to prove that A satisfies every identity of $M_n(\mathbb{Z})$. We shall do this by a series of reductions.

Let I_n denote the set of elements of $\mathbb{Z}\{x_1, x_2, \ldots\}$ that are identities for $M_n(\mathbb{Z})$. This is a T-ideal in $\mathbb{Z}\{x_1, x_2, \ldots\}$, that is, it is stabilized by all the ring endomorphisms of $\mathbb{Z}\{x_1, x_2, \ldots\}$. We shall call a ring A an I_n*-ring* if every element of I_n is an identity for A. We prove first

LEMMA 3. *If* A *is an* I_n*-ring and* A *is an algebra over* K *and* K' *is a commutative algebra over* K *then* $A_{K'} = K' \otimes_K A$ *is an* I_n*-ring.*

Proof. Since $M_n(\mathbb{Z})$ is a free $\mathbb{Z}$-module, if $mf \in I_n$ for $m \in \mathbb{Z}$ then $f \in I_n$. This implies by a Vandermonde determinant argument that if $f \in I_n$ and $f = f_0 + f_1 + \ldots + f_m \in I_n$ where f_j is homogeneous of degree j in one of the x_i then $f_j \in I_n$. Next suppose $f = f(x_1, \ldots, x_m) \in I_n$ is homogeneous of degree m_i in x_i and write

$$f(x_1, \ldots, x_{i-1}, x_i + x_{m+1}, x_{i+1}, \ldots, x_m) = \sum_{j=0}^{m_i} f_{ij}(x_1, \ldots, x_{m+1})$$

where f_{ij} is homogeneous of degree j in x_i (and of degree $m_i - j$ in x_{m+1}). Then every $f_{ij} \in I_m$. These two properties of I_n imply as in Jacobson [2], pp. 27-30, that if A/K is an I_n-ring then so is $A_{K'}$ for any commutative algebra K'/K. □

We need to prove that if A is Azumaya of rank n^2 over its center C then A is an I_n-ring. We first reduce the proof to the case in which C is noetherian by using the following

LEMMA 4. *Let* A *be an Azumaya algebra over* C. *Then there exists an Azumaya algebra* A' *over a noetherian subring* C' *of* C *such that* $A \cong C \otimes_{C'} A'$. *Moreover, if* A *has a rank then so has* A' *and their ranks are equal.*

Proof. The first assertion is proved in Orzech, Small [1]. To prove the second we write $A' = A'_1 \oplus \ldots \oplus A'_s$ where the A'_i are Azumaya over their centers of different ranks. Then $A \cong A'_C = A'_{1C} \oplus \ldots \oplus A'_{sC}$ where rank A'_{iC} = rank A_i . It follows that $s = 1$ and rank A' = rank A . □

Evidently Lemmas 3 and 4 reduce the proof that any Azumaya algebra of rank n^2 is an I_n-ring to the case in which C is noetherian. The "local-global" principal of commutative ring theory then permits a reduction to the case in which C is local noetherian. If M is the maximal ideal of the local noetherian ring C $\cap M^i = 0$ by Krull's intersection theorem. Hence we can define the completion $\hat{C}$ of C in the M-adic topology and form the algebra $\hat{A} = A_{\hat{C}}$. Since $A \hookrightarrow \hat{A}$ it suffices to prove that $\hat{A}$ is an I_n-ring. We have therefore achieved a reduction to the situation in which C is complete, local and noetherian. We now drop the noetherian condition (which was needed to obtain a completion) and we proceed to prove that if K is a complete local ring and A is Azumaya over K then A is an I_n-ring.

We recall that a ring A is called an SBI-*ring* ("suitable for building idempotents") if given any z in the Jacobson radical, rad A , there exists a $w \in$ rad A such that $w^2 - w = z$ and the centralizer $C_A(w) = C_A(z)$ (Jacobson [1], pp. 53-55). Then we have

LEMMA 5. *If* A *is an Azumaya algebra over a complete local ring* K *then* A *is an* SBI-*ring*.

Proof. If M is the maximal ideal of K then MA is a maximal ideal in A so A/MA is simple. A is a free module of finite rank over K and we have the product topology in A $(\cong K^{(n^2)})$ making A a topological ring. The completeness of K implies that if $z_i \in M^iA$ then $\sum z_i$ exists in A . In particular, if $z \in MA$ then z is quasi-regular since $1 - z$ has the inverse $1 + z + z^2 + \ldots$. Hence $MA \subset$ rad A and since A/MA is simple MA = rad A . The proof of Proposition 3 on p.54 of Jacobson [1] carries over to prove that A is SBI . □

We recall also the following useful lifting property of SBI-rings.

LEMMA 6. *Let* A *be an* SBI-*ring and let* $\{\bar{e}_{ij} \mid 1 \le i, j \le n\}$ *be a set of matrix units in* $\bar{A} = A/\text{rad}\, A$ $\left(\sum \bar{e}_{ii} = \bar{1},\ \bar{e}_{ij}\bar{e}_{kl} = \delta_{jk}\bar{e}_{il}\right)$. *Then* $\bar{e}_{ij}$ *has a lift* e_{ij} *in* A *such that* $\{e_{ij}\}$ *is a set of matrix units for* A .

This is proved in Jacobson [1], p. 55.

We require also

LEMMA 7. *Let* K *be a complete local ring with maximal ideal* M *and let* $\bar{L} = \bar{K}(\bar{\theta})$ *be a finite dimensional extension field of* $\bar{K} = K/M$. *Then there exists*

a commutative algebra L *over* K *that is a complete local ring having a* K*-rank of* $[\overline{L} : \overline{K}]$ *and such that if* N *is the maximal ideal of* L *then* $L/N \cong \overline{L}$.

Proof. Let $\overline{f}(x) \in \overline{K}[x]$ be the minimum polynomial of $\overline{\theta}$ over $\overline{K}$. Let $f(x)$ be a monic lift of $\overline{f}(x)$ in $K[x]$. Put $L = K[x]/(f(x))$. Then it is easily seen that L satisfies the conditions. □

We can now prove that any Azumaya algebra of rank n^2 over a complete local ring K is an I_n-ring. Let M be the maximal ideal of K so $MA = \text{rad } A$ and $\overline{A} = A/MA$ is central simple over the field $\overline{K} = K/M$. Let $\overline{L}$ be a separable splitting field over $\overline{K}$ of $\overline{A}$ and let L be as in Lemma 7. If we replace A by A_L we may assume $\overline{A} = A/MA = M_n(\overline{K})$. By Lemmas 5 and 6 there exists a set of matrix units $\{e_{ij} \mid 1 \le i, j \le n\}$ in A such that the $\overline{e}_{ij} = e_{ij} + MA$ form a base for $\overline{A}$ over $\overline{K}$. Then the e_{ij} form a base for A over K . Hence $A = M_n(K)$. It is now clear that A is an I_n-ring.

This completes the proof that any Azumaya algebra of rank n^2 over its center is an A_n-ring.

References

M. Artin

[1] "On Azumaya algebras and finite dimensional representations of rings", *J. Algebra* 11 (1969), 532-563.

R. Bix

[1] "Separable Jordan algebras over a commutative ring" (PhD thesis, Yale University, Connecticut, 1977). See also: *J. Algebra* (to appear).

Frank DeMeyer, Edward Ingraham

[1] *Separable Algebras Over Commutative Rings* (Lecture Notes in Mathematics, 181. Springer-Verlag, Berlin, Heidelberg, New York, 1971).

Nathan Jacobson

[1] *Structure of Rings* (American Mathematical Society Colloquium Publications 37. American Mathematical Society, Providence, Rhode Island, 1956).

[2] *Structure and Representations of Jordan Algebras* (American Mathematical Society, Colloquium Publications, 39. American Mathematical Society, Providence, Rhode Island, 1968).

[3] *PI-Algebras. An Introduction* (Lecture Notes in Mathematics, 441. Springer-Verlag, Berlin, Heidelberg, New York, 1975).

Max-Albert Knus, Manuel Ojanguren

[1] *Théorie de la Descente et Algèbres d'Azumaya* (Lecture Notes in Mathematics, 389. Springer-Verlag, Berlin, Heidelberg, New York, 1974).

Morris Orzech, Charles Small

[1] *The Brauer Group of Commutative Rings* (Marcel Dekker, New York, 1975).

Claudio Procesi

[1] "On a theorem of M. Artin", *J. Algebra* 22 (1972), 309-315.

[2] *Rings with Polynomial Identities* (Pure and Applied Mathematics, 17. Marcel Dekker, New York, 1973).

W. Schelter

[1] "On a theorem of Artin and Azumaya algebras", *J. Algebra* (to appear).

CHAPTER III

SHIRSHOV'S THEOREMS ON LOCAL FINITENESS

In this part we shall prove some theorems on local finiteness of *PI*-algebras that were proved by Shirshov in 1957. These results were overlooked by Western algebraists until quite recently and for this reason may qualify for inclusion in this series of lectures. An extension of some of Shirshov's work (dealing with alternative algebras) was published by McCrimmon in 1974. In this paper McCrimmon made use of an earlier exposition of Shirshov's work by Slater. Moreover, McCrimmon gave an account of the associative Shirshov theorems in a lecture at a Summer Institute on Ring Theory at the University of Chicago in 1973. Perhaps this was the first time that associative ring theorists in the West became aware of Shirshov's important results.

Shirshov's results generalized earlier ones by Jacobson, Levitzki and Kaplansky, especially, the local finiteness theorem of Kaplansky for *PI*-algebras that are algebraic over a field. The earlier results had been obtained using structure theory On the other hand, Shirshov's methods are elementary and combinatorial so they apply equally well to algebras over commutative rings. Shirshov also proved a local finiteness theorem for special Jordan algebras. We shall consider this also in this paper.

To state Shirshov's theorem we need to introduce some definitions. Let A be an algebra over a commutative ring K. An element $a \in A$ will be called *(integral) algebraic* if there exists a monic polynomial $f(\lambda) \in K[\lambda]$ such that $f(a) = 0$. The minimum degree for such polynomials is called the *degree* of a. If this degree is n then the subalgebra $K[a]$ generated by a is spanned over K by $1, a, \ldots, a^{n-1}$. An algebra A is called *locally finite* if the subalgebras generated by finite subsets of A are finitely spanned over K, that is, have finite sets of generators as K-modules.

The polynomial identity condition that we shall require is somewhat stronger than what is needed for some other parts of the theory, notably, for the structure theorem on prime *PI*-algebras. An element $f(x_1, \ldots, x_m) \in K\{x_1, x_2, \ldots\}$ is called *monic* if one of the monomials of highest (total) degree occurring in f has coefficient 1. We shall require that A satisfies a monic identity. It is not difficult to show, by the usual linearization process, that if A satisfies a monic identity then A satisfies a multilinear monic identity of the same degree and this can be taken to have the form

$$x_1 \ldots x_m - \sum_{\substack{\pi \neq 1 \\ \pi \in \Sigma_m}} \alpha_\pi x_{\pi 1} \ldots x_{\pi d} \tag{1}$$

where $\alpha_\pi \in K$.

We can now state

SHIRSHOV'S THEOREM. *Let* A *be an algebra over a commutative ring* K *satisfying a monic identity of degree* d. *Suppose* A *is generated by a set of elements* $\{a_i\}$ *(not necessarily finite) such that every monomial of degree less than or equal to* d *in these elements is algebraic. Then* A *is locally finite.*

The proof of this theorem is based on some combinatorial results on free monoids. We shall follow an exposition of these results due to McCrimmon which will appear in a forthcoming book of his on alternative algebras.

Let $X_n = \{x_1, x_2, \ldots, x_n\}$ and let $\mathrm{FM}(X_n)$ be the free monoid generated by X_n. Its elements are the word 1 and the words $x_{i_1} \ldots x_{i_r}$, $r \geq 1$. Equality, multiplication and degree of an element are defined as usual.

We introduce a total ordering in $\mathrm{FM}(X_n)$ by specifying that $x_1 < x_2 < \ldots < x_n$ and that $1 > u$ for any $u \neq 1$ and

$$u = x_{i_1} \ldots x_{i_r} > v = x_{j_1} \ldots x_{j_s}$$

if either u is an initial segment of v $(v = uw$, $w \neq 1)$ or

$i_1 = j_1, \ldots, i_k = j_k$ and $i_{k+1} > j_{k+1}$ for $k < \min(r, s)$. We call this the *augmented lexicographic ordering* of $FM(X_n)$.

We now list some properties of this ordering.

I. $u > v \Longleftrightarrow wu > wv$ *for any* $w \in FM(X_n)$.

Proof. Clear. □

II. *If* $u > v$ *and* u *is not an initial segment of* v *then* $uw > vw'$ *for any* w, w'.

Proof. Clear. □

We shall need to consider factorizations of a word w as $w = w_1 w_2 \ldots w_m$ where the w_i are subwords. The integer m is called the *length* of the factorization. A factorization $w_1 w_2 \ldots w_m$ is called *dominant* if

(2)
$$w > w_{\pi 1} w_{\pi 2} \ldots w_{\pi m}$$

for every $\pi \neq 1$ in Σ_m.

III. *If* $w_1 \ldots w_m$ *is dominant then* $w_r \ldots w_m$ *is a dominant factorization for every* r, $1 \leq r \leq m$.

Proof. Clear from I. □

IV. *Let* $w = w_1 \ldots w_m$ *where* $w_i = x_n^{e(i)} v_i$, $e(i) \geq 1$ *and* v_i *begins with some* $x_k \neq x_n$. *Suppose* $w = w_1 \ldots w_m$ *is dominant. Then*

(3)
$$e(1) \geq e(2) \geq \ldots \geq e(m) .$$

Proof. If $e(1) < e(2)$ then $w_2 w_1 \ldots w_m$ begins with a higher power of x_n than w, contrary to the hypothesis that $w_1 w_2 \ldots w_m$ is dominant. Thus $e(1) \geq e(2)$. Next, by III, $w_2 \ldots w_m$ is a dominant factorization. Hence $e(2) \geq e(3)$. Continuing in this way we obtain (3). □

V. *The factorization* $w = w_1 w_2 \ldots w_m$ *is dominant if each* $w_i = u_i v_i$ *where* $u_1 > u_2 > \ldots > u_m$ *and no* u_i *is an initial segment of a* u_j *with* $j > i$.

Proof. Let $\pi \neq 1$ be in Σ_m and suppose $\pi(1) = 1, \ldots, \pi(k-1) = k - 1$ but $\pi(k) \neq k$. Then $\pi(k) > k$ and $k < m$. By II,

$$w_k \ldots w_m = u_k v_k w_{k+1} \ldots w_m > u_{\pi k} v_{\pi k} w_{\pi(k+1)} \ldots w_{\pi m} .$$

Hence $w_1 \ldots w_m > w_{\pi 1} \ldots w_{\pi m}$ by I. □

Let m and M be two positive integers and let Y be the set of words of the form $x_n^e v$ where $1 \le e \le M$ and v is in the submonoid $\mathrm{FM}(X_{n-1})$ generated by $x_1, \ldots, x_{n-1}$ and $1 \le \deg v \le m$. Since the number of v of degree k in $x_1, \ldots, x_{n-1}$ is $(n-1)^k$ we see that

$$(4) \qquad |Y| = N(n, m, M) = M\big((n-1) + (n-1)^2 + \ldots + (n-1)^m\big) .$$

It is clear that the submonoid generated by Y can be identified with the free monoid $\mathrm{FM}(Y_N)$, $N = N(n, m, M)$. An element of this submonoid has a degree in the y's, called its *Y-degree* (as well as its X-degree). We order the $y_i \in Y = Y_N$ in increasing order given by the ordering $<$ in $\mathrm{FM}(X_n)$ and use this to define the augmented lexicographic order in $\mathrm{FM}(Y_N)$. We denote this order by $\ll$. We have

VI. *The order in* $\mathrm{FM}(Y_N)$ *given by* $\ll$ *coincides with the induced order given by regarding* $\mathrm{FM}(Y_N)$ *as a submonoid of* $\mathrm{FM}(X_n)$.

Proof. It suffices to show that if $u, v \in \mathrm{FM}(Y_N)$ and $u \gg v$ then $u > v$. Suppose $u \gg v$.

Case I. $v = uw$, $w \in \mathrm{FM}(Y_N)$, $w \neq 1$. Then $w \in \mathrm{FM}(X_n)$ and the result is clear.

Case II. $u = wy_i z$, $v = wy_j t$, $i > j$, $w, z, t \in \mathrm{FM}(Y_N)$.

(a) $y_i > y_j$ but y_i is not an initial segment of y_j. Then $u > v$ follows from I and II.

(b) $y_j = y_i s$, $s \in \mathrm{FM}(X_{n-1})$, $s \neq 1$. If $z = 1$ then $v = ust$ so $u > v$. If $z \neq 1$ then z begins with x_n. Hence $u = wy_i x_n \ldots$, $v = wy_i st$ so again $u > v$. □

An immediate consequence of this is that if we have a factorization of $u \in \mathrm{FM}(Y_N)$ as $u_1 \ldots u_m$ where the $u_i \in \mathrm{FM}(Y_N)$ then this is dominant in $\mathrm{FM}(Y_N)$ if and only if it is dominant in $\mathrm{FM}(X_n)$.

We are now ready to prove the

FIRST COMBINATORIAL LEMMA. *For given positive integers* n, m, M *there exists a positive integer* $f(n, m, M)$ *such that any word* w *in* $\mathrm{FM}(X_n)$ *with* $\deg w > f(n, m, M)$ *contains a subword of one of the following two forms:*

(i) $w_0 = u^M$, $\deg u \ge 1$;

(ii) w_0 *has a dominant factorization* $w_0 = u_1u_2 \ldots u_m$ *of length* m .

Proof. We use induction on m . We can start the induction with $f(n, 1, M) = 1$ since any x_i is dominant of length 1 . Also for a given m we can use a sub-induction on n . Here we can start with $f(1, m, M) = M$ since any word in x_1 of degree greater than M contains the subword x_1^M . Assume we have defined $f(n, m-1, M)$ for all n and $f(n-1, m, M)$ to satisfy the conditions for $(n, m-1, M)$ and $(n-1, m, M)$. Put

(5) $$f(n, m, M) = \left(M+f(n-1, m, M)\right)\left(2+f(N, m-1, M)\right)$$

where $N = N\left(n, f(n-1, m, M), M\right)$ as defined in (4).

Let w be a word in $x_1, \ldots, x_n$ of degree greater than $f(n, m, M)$ and write

(6) $$w = \left\{x_n^{e(0)}v_0\right\}\left\{x_n^{e(1)}v_1\right\} \ldots \left\{x_n^{e(r+1)}v_{r+1}\right\}$$

where $v_i \in \mathrm{FM}\left(X_{n-1}\right)$, $e(0) \geq 0$, $e(i) > 0$ if $i > 0$, $\deg v_i > 0$ for $0 \leq i \leq r$. Then

(7) $$w = \left\{x_n^{e(0)}v_0\right\}w'\left\{x_n^{e(r+1)}v_{r+1}\right\} .$$

If some $e(i)$ in (6) is greater than M then we have the subword x_n^M and we are done. Also if $\deg v_i > f(n-1, m, M)$ then the induction on n implies that we have a subword w_0 of v_i satisfying *(i)* or *(ii)* so again we are done. Hence we may assume every $e(i) \leq M$ and every $\deg v_i \leq f(n-1, m, M)$. Then

$$\begin{aligned}\deg w &\leq (r+2)M + (r+2)f(n-1, m, M) \\ &= (r+2)\left(M+f(n-1, m, M)\right) .\end{aligned}$$

On the other hand,

$$\deg w > f(n, m, M) = \left(M+f(n-1, m, M)\right)\left(2+f(N, m-1, M)\right) .$$

Hence

(8) $$f(N, m-1, M) < r .$$

Now consider the set of words $x_n^e v$, $1 \leq e \leq M$, $v \in \mathrm{FM}\left(X_{n-1}\right)$, $1 \leq \deg v \leq f(n-1, m, M)$. The number of these words is $N = N\left(n, f(n-1, m, M), M\right)$ as in (4). The word w' in (7) is $\left\{x_n^{e(1)}v_1\right\} \ldots \left\{x_n^{e(r)}v_r\right\}$ so its Y-degree is $r > f(N, m-1, M)$. Hence by induction on m , w' contains a subword $w_0' = u'^M$, $\deg u' \geq 1$, or a word w_0' with a dominant factorization $u_1' \ldots u_{m-1}'$ in $\mathrm{FM}\left(Y_N\right)$ and hence in $FM(X_n)$ of length $m - 1$. In the first case we are done. Hence we assume the second and we proceed to expand $u'_1 \dots u'_{m-1}$ to a dominant subword $w_0 = u_1 \dots u_m$ of w.

Taking into account the definition of Y_N we see that $u'_i = x_n^{e'(i)} v'_i$ where $e'(i) \geq 1$ and v'_i begins with an $x_{k_i} \neq x_n$. Then, by IV, we have

$$(9) \qquad e'(1) \geq e'(2) \geq \dots \geq e'(m-1) .$$

Since $e(r+1) \geq 1$ there is an x_n in w after u'_{m-1}. Hence

$$(10) \qquad w = pu'_1 \dots u'_{m-1} q x_n s$$

where q does not involve x_n. We claim that $u'_1 \dots u'_{m-2}(u'_{m-1}q)$ is also a dominant factorization. Let $\pi \neq 1$ be in Σ_{m-1}. Suppose first that π fixes $m - 1$. Then applying π to the factorization $u'_1 \dots u'_{m-2}(u'_{m-1}q)$ gives $u'_{\pi 1} \dots u'_{\pi(m-2)}(u'_{m-1}q) < u'_1 \dots u'_{m-2}(u'_{m-1}q)$ since $u'_1 \dots u'_{m-1}$ is dominant (using II). Next suppose π moves $m - 1$. Then applying π to the factorization $u'_1 \dots u'_{m-2}(u'_{m-1}q)$ gives the factorization

$$u'_{\pi 1} \dots (u'_{m-1}q) u'_{\pi j} \dots < u'_{\pi 1} \dots u'_{m-1} u'_{\pi j} \dots q$$

(since $u'_{\pi j}$ begins with x_n and q does not). But

$$u'_{\pi 1} \dots u'_{m-1} u'_{\pi j} \dots q < u'_1 \dots u'_{m-1} q$$

by II and the dominance of the factorization $u'_1 \dots u'_{m-1}$. Thus $u'_1 \dots u'_{m-2}(u'_{m-1}q)$ is dominant and so if we replace u'_{m-1} by $u'_{m-1}q$ we may assume $q = 1$ and we have the subword $u'_1 \dots u'_{m-1} x_n$ in w.

Now $u'_i = x_n^{e'(i)} v'_i$ where $e'(i) \geq 1$, v'_i begins with an $x_{k_i} \neq x_n$ and (9) holds. Then let

$$(11) \quad w_0 = x_n \left\{ x_n^{e'(1)-1} v'_1 x_n \right\} \left\{ x_n^{e'(2)-1} v'_2 x_n \right\} \dots \left\{ x_n^{e'(m-1)-1} v'_{m-1} x_n \right\} = u_1 u_2 \dots u_m$$

where $u_1 = x_n$, $u_2 = x_n^{e'(1)-1} v'_1 x_n$, ..., $u_m = x_m^{e'(m-1)-1} v'_{m-1} x_n$. We claim that $u_1 u_2 \dots u_m$ is a dominant factorization of w_0. Let $\pi \neq 1$ be in Σ_m. Suppose first that $\pi 1 = 1$. Then

$$u_{\pi 1} \dots u_{\pi m} = x_n \left\{ x_n^{e'(\sigma 1)-1} v'_{\sigma 1} x_n \right\} \left\{ x_n^{e'(\sigma 2)-1} v'_{\sigma 2} x_n \right\} \dots \left\{ x_n^{e'(\sigma(m-1))-1} v'_{\sigma(m-1)} x_n \right\}$$

where $\sigma \neq 1$ in Σ_{m-1}. Then

$$u_{\pi 1} \dots u_{\pi m} = u'_{\sigma 1} \dots u'_{\sigma(m-1)} x_n < u'_1 \dots u'_{m-1} x_n = u_1 \dots u_m$$

by the dominance of the factorization $u'_1 \dots u'_{m-1}$. Next suppose $\pi 1 \neq 1$. Then applying π gives $x_n^{e'(j)-1} v'_j x_n \dots < u_1 \dots u_m$ since $u_1 \dots u_m$ begins with $x_n^{e'(1)}$ and $e'_1(1) \geq e'(j) > e'(j) - 1$. Thus $w_0 = u_1 \dots u_m$ is dominant. □

We need to improve the first lemma to the

SECOND COMBINATORIAL LEMMA. *If* n, m, M *are positive integers there exists a positive integer* $g(n, m, M)$ *such that any word* w *in* $\mathrm{FM}(X_n)$ *of degree greater than* $g(n, m, M)$ *contains a subword* w_0 *having one of the following forms:*

(i) $w_0 = u^M$, $1 \leq \deg u \leq m$;

(ii) w_0 *has a dominant factorization of length* m.

The improvement over the first lemma is that we are able to require that $\deg u \leq m$ in case *(i)*. The proof will be based on the following

SUBLEMMA. *Let* l *and* m *be positive integers with* $l > m$ *and let* u *be a word of degree* l. *Then either* $u = v^e$ *for a word* v *and* $e \neq 1$ *a divisor of* l *or* u^{2m} *contains a subword that has a dominant factorization of length* m.

Proof. Write $u = z_1 \dots z_l$ where the $z_j \in X_n$ and let $\sigma = (12 \dots l)$. Let H be the subgroup of $\langle\sigma\rangle$ of $\tau = \sigma^k$ such that $z_{\tau 1} \dots z_{\tau l} = z_1 \dots z_l$. Then $H = \langle\sigma^d\rangle$ for $1 \leq d \leq l$ and $d|l$, say $l = de$. Then $u = z_1 \dots z_l = z_{d+1} \dots z_{d+l}$ (indices reduced mod l). If $d \neq l$ then $e > 1$ and $u = v^e$, $v = z_1 \dots z_d$. Next suppose $d = l$. Then the words $\sigma^k u \equiv z_{\sigma^k 1} \dots z_{\sigma^k l}$, $1 \leq k \leq l$, are distinct so we have a permutation π of $1, 2, \dots, l$ such that

$$\sigma^{\pi 1} u > \sigma^{\pi 2} u > \dots > \sigma^{\pi l} u .$$

Consider the element u^2. This can be written as

$$u^2 = z_1 \dots z_l z_1 \dots z_l = z_1(\sigma u) z_2 \dots z_l = z_1 z_2 (\sigma^2 u) z_3 \dots z_l =$$

$$= \dots = z_1 \dots z_k (\sigma^k u) z_{k+1} \dots z_l = \dots .$$

Hence for any k, $1 \leq k \leq l$, $u^2 = v_k (\sigma^k u) v'_k$ where $v_k = z_1 \dots z_k$ and $v'_k = z_{k+1} \dots z_l$. Since $l > m$,

$$u^{2m} = (u^2)^m = \left\{v_{\pi 1}(\sigma^{\pi 1}u)v'_{\pi 1}\right\} \dots \left\{v_{\pi m}(\sigma^{\pi m}u)v'_{\pi m}\right\}$$
$$= v_{\pi 1}u_1 \dots u_m$$

where

$$u_i = (\sigma^{\pi i}u)v'_{\pi i}v_{\pi(i+1)} \quad (1 \le i < m) ,$$

and

$$u_m = (\sigma^{\pi m}u)v'_{\pi m} .$$

Since the $\sigma^{\pi i}u$ all have the same length and $\sigma^{\pi 1}u > \sigma^{\pi 2}u > \dots$ it follows from V that $u_1 \dots u_m$ is a dominant factorization and $u_1 \dots u_m$ is a subword of u^{2m} . □

We now give the

Proof of the Second Combinatorial Lemma. Let $f(n, m, M)$ be as in the first lemma and put $g(n, m, M) = f(n, m, \overline{M})$ where $\overline{M} = \max(2m, M)$. Let w be a word of degree greater than $g(n, m, M) = f(n, m, \overline{M})$. By the first lemma, either w contains a subword $w_0 = u^{\overline{M}}$ with $\deg u \ge 1$ or a word w_0 with a dominant factorization of length m . We are done in the second case and also in the first case if $\deg u \le m$. Hence suppose $\deg u = l > m$. We claim that $u^{\overline{M}}$ either contains a subword u_0^M with $1 \le \deg u_0 \le m$ or a subword with a dominant factorization of length m . We prove this by induction on l . By the sublemma, either $u = v^e$ where $e | l$ and $e \ne 1$ or u^{2m} contains a subword having a dominant factorization of length m . In the first case, $\deg v < l$ and $u^{\overline{M}}$ contains the subword $v^{\overline{M}}$. Then the result follows by the degree induction. In the second case, u^{2m} and hence $u^{\overline{M}}$ contains a subword with a dominant factorization of length m . □

We are now ready to give the

Proof of Shirshov's Theorem. We assume first that $|\{a_i\}| = n < \infty$ and we shall prove that A is finitely generated as a K-module. Consider the free monoid $\mathrm{FM}(X_n)$ and the free algebra $K\{X_n\}$ having $\mathrm{FM}(X_n)$ as base over K with multiplication determined by that in $\mathrm{FM}(X_n)$. Let η be the homomorphism of $K\{X_n\}$ into A such that $x_i \mapsto a_i$, $1 \le i \le n$, and let $I = \ker \eta$. The subset U of monomials in the x's of positive degree less than or equal to d is finite. Hence there exists a positive integer e such that for every $u \in U$ there exists a monic polynomial in $K[\lambda]$ of degree e such that $f(\eta u) = 0$. Let $g(n, d, e)$ be as in the second combinatorial lemma and let V be the set of monomials in the x's of degree less than or equal to $g = g(n, d, e)$. We claim that every monomial in the x's of degree greater than g is congruent modulo I to a linear combination of monomials in V . We use induction on the degree of the monomial and for a given degree induction on the order as defined in $\mathrm{FM}(X_n)$. For a given degree r the first monomial (in the ordering) is x_1^r and if $r > g$ then $r > e$ (since the proofs of the combinatorial lemmas show that $g \geq e$ if $d \neq 1$). Now a_1^e is a linear combination with coefficients in K of $1, a_1, \ldots, a_1^{e-1}$. It follows that x_1^r is congruent modulo I to a linear combination of monomials of degree less than r . Then the result follows for x_1^r by the degree induction. Now let w be any monomial in the x's of degree greater than g . By the second combinatorial lemma either w has a factor of the form u^e , $u \in U$, or it has a monomial factor w_0 that has a dominant factorization $u_1 \ldots u_d$ of length d . In the first case the argument we used for x_1^r shows that w is congruent modulo I to a linear combination of elements in the set V . In the second case we use the fact that A satisfies an identity of the form

$$x_1 \ldots x_d - \sum_{\pi \neq 1} \alpha_\pi x_{\pi 1} \ldots x_{\pi d}$$

to conclude that $u_1 \ldots u_d$ is congruent modulo I to a linear combination of monomials of the same degree and lower order. It follows that w is congruent modulo I to a K-linear combination of monomials of the same degree and of lower order. Hence w is congruent to a linear combination of monomials contained in the set V . Since this set is finite it follows that any monomial in $a_1, \ldots, a_n$, and hence any element of A is a K-linear combination of a finite subset of A . Thus A is finitely generated as K-module.

To finish the proof we suppose $\{a_i\}$ is any set of generators satisfying the hypothesis and we let $\{b_1, \ldots, b_r\}$ be a finite subset of A . We have to show that the subalgebra B generated by $b_1, \ldots, b_r$ is finitely generated as K-module. Now the b_j are contained in a subalgebra A' generated by a finite subset of the set $\{a_i\}$, say, $\{a_1, \ldots, a_n\}$. By what we have proved, A' is finitely generated as K-module. Hence the result required will follow from the following:

LEMMA. *Let* A *be an algebra that is finitely generated as* K*-module,* B *a* *subalgebra that is finitely generated as K-algebra. Then* B *is finitely generated as K-module.*

Proof. Let $\{u_1, \ldots, u_n\}$ be a set of generators of A as K-module and let $\{b_1, \ldots, b_m\}$ be a set of generators of B as K-algebra. We have

$$u_i u_j = \sum \gamma_{ijk} u_k , \quad \gamma_{ijk} \in K ,$$

$$1 = \sum \gamma_i u_i , \quad \gamma_i \in K ,$$

$$b_l = \sum \mu_{li} u_i , \quad \mu_{li} \in K .$$

Let K' be the subring of K generated by the finite set $\{\gamma_{ijk}, \gamma_i, \mu_{li}\}$. Then K' is noetherian and $A' = \sum K' u_i$ is a K'-subalgebra of K containing the K'-subalgebra B' generated by the b_l. Since K' is noetherian B' is finitely generated as K'-module by a subset, say, $v_1, \ldots, v_r$. Then every monomial in the b_l is a K'-linear combination of the v's and hence every element of B is a K-linear combination of the v's. Thus the v's form a set of generators for B as K-module. □

This completes the proof of Shirshov's Theorem. □

Shirshov's Theorem can be carried over to *PI*-algebras without unit. Moreover, the proof gives a stronger result in the case of nil algebras. To state this we require the concepts of nilpotency and local nilpotency for algebras. An algebra A is called *nilpotent* if there exists an integer s such that $A^s = 0$, which is equivalent to saying that the product of any s elements of A is 0. A is called *locally nilpotent* if the subalgebras generated by finite subsets are nilpotent. It is an old result of Amitsur's that any nil *PI*-algebra is locally nilpotent. The following stronger result is due to Shirshov.

THEOREM 1. *Let* A *be an algebra without unit over a commutative ring* K *satisfying a monic identity of degree* d. *Suppose also that* A *is generated by a subset* $\{a_i\}$ *such that every monomial of degree less than or equal to* d *in the* a_i *is nilpotent. Then* A *is locally nilpotent.*

Proof. The proof is similar but somewhat simpler than the proof of the general local finiteness theorem. First, let $\{a_i\} = \{a_1, a_2, \ldots, a_n\}$. Then there exists an e such that $b^e = 0$ for every monomial in the a_i of degree less than or equal to d. Let $g = g(n, d, e)$ as in the Second Combinatorial Lemma. Then we claim that any product of $g' = g + 1$ elements of A is 0, so $A^{g+1} = 0$. It suffices to show this for all products of g' a_i's. As before, let η be the homomorphism of $K\{X_n\}'$ the free associative algebra without unit on the n generators $x_1, x_2, \ldots, x_n$ such that $x_i \mapsto a_i$, $1 \le i \le n$, and let $I = \ker \eta$. If u is any monomial of degree less than or equal to d in the x_i then $u^e \in I$. Now consider the set V of monomials in the x's of degree g'. The first of these in the augumented lexicographic ordering is $x_1^{g'}$ and this is in I since $g' \ge e$. Now consider the monomial $z_1 \ldots z_{g'}$, $z_i \in X_n$. By the second combinatorial lemma, $z_1 \ldots z_{g'}$ either contains a subword u^e with $1 \le \deg u \le d$, or it contains a subword with a dominant factorization $u_1 \ldots u_d$ of length d. In the first case $u^e \in I$ and hence $z_1 \ldots z_{g'} \in I$. In the second case $u_1 \ldots u_d$ is congruent modulo I to a K-linear combination of monomials of lower order and the same degree. In this case induction on the order implies that $z_1 \ldots z_{g'} \in I$. Hence the product of any g' elements a_i is 0. The proof for arbitrary sets $\{a_i\}$ is an immediate consequence of the result for finite sets of generators. □

It is a well known result that is easily proved that if A is algebraic of bounded degree then A satisfies a monic identity (Jacobson [2], p. 14). The same is true of algebras without unit and, in particular, of nil algebras. We therefore have the following consequences of the foregoing results.

COROLLARY 1. *If A is an algebra over a commutative ring K and A is algebraic of bounded degree then A is locally finite.*

COROLLARY 2. *If A is a nil algebra over K of bounded degree then A is locally nilpotent.*

We shall consider next Shirshov's local finiteness theorem for special Jordan algebras. Again let A be an associative algebra over a commutative ring K. A *special Jordan algebra* J *in* A is a K-submodule of A containing 1 and closed under the binary product aba. Since $a^2 = a1a$, $a^3 = aaa$, $a^{n+2} = aa^na$ it follows that A is closed under the unary compositions $a \mapsto a^n$. It is easily seen also that A is closed under the trilinear product $\{abc\} = abc + cba$ and the bilinear product $a \circ b = \{a1b\} = ab + ba$. A itself is a special Jordan algebra in A. We denote this as A^+. More interesting examples are obtained from associative algebras with involution (A, j). Then the subset $H(A, j)$ of j-symmetric elements of A is a Jordan algebra in A. Homomorphisms of special Jordan algebras are defined to be K-module homomorphisms η such that $\eta 1 = 1$ and $\eta(aba) = (\eta a)(\eta b)(\eta a)$.

If J is a special Jordan algebra in A we let Env J denote the (associative) subalgebra of A generated by J. We have the following

PROPOSITION 1. *If* J *is finitely generated as* K*-module then so is* Env J. *Conversely, if* Env J *is finitely generated as* K*-module and* J *is finitely generated as Jordan algebra then* J *is finitely generated as* K*-module.*

Proof. Let $\{u_1, \ldots, u_n\}$ be a subset of J such that $J = \sum K u_i$. Evidently the u_i generate Env J as algebra so every element of Env J is a linear combination of the monomials $u_{i_1} \ldots u_{i_r}$. We claim that every element of A is a linear combination of 1 and the monomials $u_{i_1} \ldots u_{i_r}$ in which the i_j are distinct. Since the number of these is finite this will prove the first statement. It suffices to show that every $u_{j_1} \ldots u_{j_s}$ in which $j_k = j_l$ for some $l > k$ can be expressed as a linear combination of $u_{i_1} \ldots u_{i_r}$ with distinct i_j. We prove this by induction on s and on $l - k$. If $l = k + 1$,

$$u_{j_1} \ldots u_{j_s} = u_{j_1} \ldots u_{j_k}^2 \ldots u_{j_s}$$

and since $u_{j_k}^2 \in J$ we have $u_{j_k}^2 = \sum \alpha_{kl} u_l$. Substituting this gives an expression for $u_{j_1} \ldots u_{j_s}$ as a linear combination of monomials that are products of $s - 1$ u's. Then we can invoke the degree induction. Next let $l - k > 1$. Then we can use the relation $u_{j_k} u_{j_{k+1}} + u_{j_{k+1}} u_{j_k} = u_{j_k} \circ u_{j_{k+1}} = \sum \beta_l u_l$ to replace $u_{j_k} u_{j_{k+1}}$ by $-u_{j_{k+1}} u_{j_k} + \sum \beta_l u_l$. This gives an expression for $u_{j_1} \ldots u_{j_s}$ as a linear combination of monomials to which the induction applies. This proves the first statement.

The second statement is a consequence of the following result: if J' is a special Jordan algebra that is finitely generated as K-module and J is a subalgebra that is finitely generated as Jordan algebra then J is finitely generated as K-module. The proof of this is identical with that of the lemma in the proof of Shirshov's Theorem. The result we require is obtained by taking $J' = \text{Env } J$. □

The most natural way of defining identities for special Jordan algebras is to first define free special Jordan algebras. If $X = \{x_1, x_2, \ldots\}$ we can define the *free special Jordan algebra* FSJ(X) over K to be the subalgebra of $K\{X\}^+$ generated by X. In other words, this is the smallest K-submodule of $K\{X\}$

containing 1 and X and closed under the product aba. It is easily seen that FSJ(X) has the freeness property that any map of X into a special Jordan algebra J can be extended in one and only one way to a homomorphism of FSJ(X) into J. In a similar manner we can define the free special Jordan algebra $\mathrm{FSJ}(X_n) \subset K\{X_n\}$ where $X_n = (x_1, \ldots, x_n)$. The elements of FSJ(X) or $\mathrm{FSJ}(X_n)$ are called *Jordan polynomials* or *Jordan elements* of $K\{X\}$ or $K\{X_n\}$.

If J is a special Jordan algebra in the associative algebra A over K, an element $g(x_1, \ldots, x_m) \in \mathrm{FSJ}(X)$ is called an *identity* for J if g is mapped into 0 by every homomorphism of FSJ(X) into J. If we denote the image of g under the homomorphism such that $x_i \mapsto a_i$, $1 \le i \le m$, by $g(a_1, \ldots, a_m)$ then g is an identity for J if and only if $g(a_1, \ldots, a_m) = 0$ for all $a_i \in J$. g is called *monic* if it is a monic element of $K\{X\}$.

Associative ring theorists have considered a somewhat different concept of identity for the case of the special Jordan algebra of symmetric elements of an associative algebra with involution. They have considered arbitrary elements $f(x_1, \ldots, x_m) \in K\{X\}$ and required that $f(a_1, \ldots, a_m) = 0$ for all $a_i \in J$. This is equivalent to: f if mapped into 0 by every homomorphism of $K\{X\}$ into A such that $x_i \mapsto a_i \in J$, $1 \le i \le m$. We shall call an element of this sort an *associative identity* for J.

It turns out that it does not matter which of these notions we use, for, as we shall show, a special Jordan algebra has a monic identity if and only if it has an associative monic identity. For the proof of this we need to look at the elements of FSJ(X). We recall first that $K\{X\}$ (or $K\{X_n\}$) has a unique involution ρ such that $x_{i_1} \ldots x_{i_r} \mapsto x_{i_r} \ldots x_{i_1}$. This is called the *reversal involution*. Let $H(K\{X\}, \rho)$ denote the subset of $K\{X\}$ of symmetric elements under ρ. This is a special Jordan algebra in $K\{X\}$ containing X. Hence $\mathrm{FSJ}(X) \subset H(K\{X\}, \rho)$. Similarly, $\mathrm{FSJ}(X_n) \subset H(K\{X_n\}, \rho)$. If $n = 2$ we have

PROPOSITION 2. $\mathrm{FSJ}(X_2) = H(K\{X_2\}, \rho)$.

Proof. Write $x = x_1$, $y = x_2$. Any ρ-symmetric element of $K\{x, y\}$ is a linear combination of elements of the following forms:

$$\ldots x^{i_2}y^{j_1}x^{i_1}y^{j_1}x^{i_2} \ldots$$

$$\ldots y^{j_2}x^{i_1}y^{j_1}x^{i_1}y^{j_2} \ldots$$

$$x^{i_1}y^{j_1} \ldots x^{i_r} + x^{i_r} \ldots y^{j_1}x^{i_1}$$

$$y^{j_1}x^{i_1} \ldots y^{j_r} + y^{j_r} \ldots x^{i_1}y^{j_1}$$

$$x^{i_1}y^{j_1} \ldots x^{i_r}y^{j_r} + y^{j_r}x^{i_r} \ldots y^{j_1}x^{i_1}, \quad i_k, j_k > 0 .$$

Those in the first two lines are clearly Jordan polynomials and if we can show that the ones in the last line are Jordan polynomials it will follow that those in the third and fourth lines are also Jordan polynomials. For the ones in the last line we prove the result by induction on the *height* r. We have

$$x^{i_1}(y^{j_1}x^{i_2} \ldots x^{i_r} + x^{i_r} \ldots x^{i_2}y^{j_1})y^{j_r} + y^{j_r}(y^{j_1}x^{i_2} \ldots x^{i_r} + x^{i_r} \ldots x^{i_2}y^{j_1})x^{i_1}$$
$$= (x^{i_1}y^{j_1} \ldots x^{i_r}y^{j_r} + y^{j_r}x^{i_r} \ldots y^{j_1}x^{i_1})$$
$$+ (x^{i_1+i_r}y^{j_{r-1}} \ldots x^{i_2}y^{j_1+j_r} + y^{j_1+j_r}x^{i_2} \ldots y^{j_{r-1}}x^{i_1+i_r}) .$$

The height induction implies that the left hand side and the second parenthesis on the right hand side are Jordan polynomials. It follows that

$$x^{i_1}y^{j_1} \ldots x^{i_r}y^{j_r} + y^{j_r}x^{i_r} \ldots y^{j_1}x^{i_1}$$

is a Jordan polynomial. □

We can now prove

PROPOSITION 3. *J has a monic identity if and only if it has an associative monic identity.*

Proof. Since a monic identity is an associative monic identity it remains to show that if J has an associative monic identity then it has a monic identity. By linearization we can show that if J has an associative monic identity then it has a multilinear one, say,

$$f = x_1 \ldots x_n + \sum_{\pi \neq 1} \alpha_\pi x_{\pi 1} \ldots x_{\pi n} .$$

Now apply the homomorphism of $K\{X\}$ into $K\{x, y\}$ such that $x_i \mapsto xy^i$, $i = 1, 2, \ldots$. The image of f under this homomorphism is the monic homogeneous polynomial

$$f(xy, xy^2, \ldots, xy^n) = xyxy^2 \ldots xy^n + \ldots .$$

Next apply the reversal operator ρ and form

$$g(x, y) = f(xy, \ldots, xy^m)\rho f(xy, xy^2, \ldots) = xyxy^2 \ldots xy^{2n}x \ldots y^2xyx + \ldots .$$

This is monic and symmetric and is an identity for J. By Proposition 2, $g \in \mathrm{FSJ}(x, y)$.

We have seen that any special Jordan algebra is closed under powers. We can therefore define algebraic elements as in the associative case: $a \in J$ is *algebraic* if there exists a monic $f(\lambda) \in K[\lambda]$ such that $f(a) = 0$. The least degree for such polynomials is called the *degree* of a.

To state the local finiteness theorem for special Jordan algebras we require also the concept of *Jordan monomial* of $\mathrm{FSJ}(X)$. We define these inductively by: 1 and the x's are Jordan monomials and if p, q and r are Jordan monomials then so are pqp and $pqr + rqp$. It is clear from the definition that Jordan monomials are homogeneous elements of $K\{X\}$. We note also that if p is a Jordan monomial then so is p^k for any $k \geq 0$ and if p and q are Jordan monomials then so is $p \circ q = pq + qp = p1q + q1p$. Since

$$\left(\sum \alpha_i p_i\right) q \left(\sum \alpha_i p_i\right) = \sum \alpha_i^2 p_i q p_i + \sum_{i<j} \alpha_i \alpha_j \left(p_i q p_j + p_j q p_i\right)$$

it is clear that the set of K-linear combinations of the Jordan monomials is a subalgebra of $\mathrm{FSJ}(X)$ containing X. Hence it coincides with $\mathrm{FSJ}(X)$. Thus any Jordan polynomial is a linear combination of Jordan monomials.

If $a_1, \ldots, a_n$ are elements of a special Jordan algebra and $p(x_1, x_2, \ldots, x_n)$ is a Jordan monomial then $p(a_1, \ldots, a_n)$ will be called a *Jordan monomial in the* a_i.

We are now ready to state an extension of Shirshov's theorem on special Jordan algebras.

THEOREM 2. *Let* J *be a special Jordan algebra in the associative algebra* A *and assume* $\mathrm{Env}\, J = A$. *Suppose* J *satisfies a monic identity and* J *has a set of generators* $\{a_i\}$ *(not necessarily finite) such that every Jordan monomial in the* a_i *is algebraic. Then* A *and* J *are locally finite.*

We remark that the hypothesis $\mathrm{Env}\, J = A$ is not a real restriction since it can be achieved by replacing A by $\mathrm{Env}\, J$.

For the proof of the theorem we shall require the concept of a k-word in $\mathrm{FM}(X_n)$ and another combinational lemma on free monoids. If $1 \leq k \leq n$ we define a *k-word* as either a power x_k^e, $e > 1$, or a word $x_k w x_i$ where $w \in \mathrm{FM}(X_k)$ and $i < k$. The k-words of the first kind are Jordan monomials and any k-word of the second kind is the highest word in a suitable Jordan monomial. More precisely, if

$x_k w x_i$ is a k-word then there exists a Jordan monomial q in FSJ(X) such that

$$q = x_k w x_i + \sum \alpha_{w'} w'$$

where $w' \in \mathrm{FM}(X_k)$ and w' begins with some x_j, $j < k$. Then $w' < x_k w x_i$. To see this we write

$$x_k w x_i = x_k^{e_1} w_1 x_k^{e_2} w_2 \ldots x_k^{e_r} w_r$$

where the $e_i \geq 1$, $w_i \in \mathrm{FM}(X_{k-1})$ and $\deg w_i \geq 1$. Suppose first that $r = 1$ and write $w_1 = x_{i_1} \ldots x_{i_s}$, $1 \leq i_j \leq k-1$, $s \geq 1$. Consider the Jordan monomial

$$q_1 = \left(\ldots\left(\left(x_k^{e_1} \circ x_{i_1}\right) \circ x_{i_2}\right) \ldots\right) \circ x_{i_s} .$$

It is clear that the only term beginning with x_k in q_1 is $w = x_k^{e_1} x_{i_1} \ldots x_{i_s}$.

Now suppose we have a Jordan monomial $q_{r-1} \in \mathrm{FM}(X_k)$ such that the only term beginning with x_k in q_{r-1} is $x_k^{e_1} w_1 \ldots x_k^{e_{r-1}} w_{r-1}$ and let $w_r = x_{i_1} \ldots x_{i_s}$. Put

$$q_r = \begin{cases} q_{r-1} x_k^{e_r} x_{i_1} + x_{i_1} x_k^{e_r} q_{r-1} & \text{if } s = 1 , \\ \left(\ldots\left(\left(q_{r-1} x_k^{e_r} x_{i_1} + x_{i_1} x_k^{e_r} q_{r-1}\right) \circ x_{i_2}\right) \ldots\right) \circ x_{i_s} & \text{if } s > 1 . \end{cases}$$

Then it is readily seen that the only term beginning with x_k in q_r is the given k-word $w = x_k^{e_1} w_1 \ldots x_k^{e_r} w_r$.

The combinatorial lemma we require is the

THIRD COMBINATORIAL LEMMA. *Let m, n be positive integers, $M(n)$ a positive integral valued function of n. Then there exists a positive integer $d(n, m, M(n))$ such that any word $w \in \mathrm{FM}(X_n)$ of degree greater than $d(n, m, M(n))$ contains a subword w_0 such that for some k, $1 \leq k \leq m$, either*

(i) $w_0 = x_k^{M(1)}$ *; or*

(ii) $w_0 = u^{M(k)}$ *for a k-word u such that*

$$1 \le \deg u \le \delta(k) = m\big(M(1)+d(k-1,\ m,\ M(k-1))\big)$$

where $\delta(1) = mM(1)$; *or*

(iii) w_0 *has a dominant factorization* $w_0 = u_1 \ldots u_m$ *of length* m *in which every* u_i *is a* k*-word.*

We delay the proof of this lemma and use it to give the

Proof of Theorem 2. We assume first that $\{a_i\} = \{a_1, \ldots, a_n\}$ and that we have a polynomial

$$(12) \qquad f(x_1, \ldots, x_m) = x_1 \ldots x_m - \sum_{\pi \ne 1} \alpha_\pi x_{\pi 1} \ldots x_{\pi m}$$

such that $f(b_1, \ldots, b_m) = 0$ for all $b_j \in J$. Let $M(1)$ be the maximum degree of algebraicity of the a_i and define $M(k)$ inductively as the maximum degree of algebraicity of the Jordan monomials in the a_i of degree (as given in the free algebra) at most $\delta(k) = m\big(M(1) + d(k-1, m, M(k-1))\big)$ where $d(k-1, m, M(k-1))$ is as defined in the Third Combinatorial Lemma. Since the a_i generate J and J generates A, the a_i generate A. We have a homomorphism η of $K\{X_n\}$ into A such that $x_i \mapsto a_i$, $1 \le i \le n$. Let $I = \ker \eta$. To prove that A is finitely generated as K-module it suffices to show that every $w \in \mathrm{FM}(X_n)$ of degree greater than $d = d(n, m, M(n))$ is congruent modulo I to a linear combination of elements of $\mathrm{FM}(X_n)$ of degree less than or equal to d. We use induction on the degree of w and for a given degree induction on the order. We apply the Third Combinatorial Lemma to w and the subword w_0. In case *(i)* we use the fact that $a_k^{M(1)}$ is a linear combination of $1, a_k, \ldots, a_k^{M(1)-1}$ to conclude that w is congruent modulo I to a linear combination of monomials of lower degree. Then the degree induction is applicable. In case *(ii)* we write $u = q + u'$ where q is a Jordan monomial of the same degree as u and u' is a linear combination of elements of $\mathrm{FM}(X_k)$ of the same degree as u and of lower order than u. Then $u^{M(k)} = q^{M(k)} + u''$ where u'' is a linear combination of monomials of the same degree and lower order than $u^{M(k)}$. Since $q^{M(k)}$ is congruent modulo I to a linear combination of $1, q, \ldots, q^{M(k)-1}$, degree induction and order induction give the result in this case. Now suppose we have case *(iii)*: $w_0 = u_1 \ldots u_m$ where the u_i are k-words and the factorization is dominant. We write $u_i = q_i + u_i'$ where q_i is a Jordan monomial and u_i' is a linear combination of monomials of the same degree as u_i and of lower order. Then, by (12) applied to the image of q_i, we have

$$u_1 \dots u_m = q_1 \dots q_m + u'' \equiv \sum_{\pi \neq 1} \alpha_\pi q_{\pi 1} \dots q_{\pi m} + u'' \equiv \sum \alpha_\pi u_{\pi 1} \dots u_{\pi m} + u''' \pmod{J}$$

where u''' is a linear combination of monomials of lower order than $u_1 \dots u_m$. Since $u_1 \dots u_m$ is a dominant factorization the result follows in this case by the order induction.

Now suppose $\{a_i\}$ is arbitrary and let $b_1, \dots, b_l$ be arbitrary elements of A. Then these elements are contained in a subalgebra of A generated by a finite subset of $\{a_i\}$, say, $\{a_1, \dots, a_n\}$. By what we have proved, the subalgebra generated by the a_i is finitely generated as K-module. Hence, as we showed before, the subalgebra generated by the b_j is a finitely generated K-module. This shows that A is locally finite. It remains to show that J is locally finite. To see this we now suppose that the $b_j \in J$. Then the subalgebra of A generated by the b_j is finitely generated as K-module. Hence, by Proposition 1, the subalgebra of J generated by the b_j is a finitely generated K-module. □

It remains to give the

Proof of the Third Combinatorial Lemma. For any n, m, M let $g(n, m, M)$ be as in the Second Combinatorial Lemma. Define $d(n, m, M(n))$ inductively by

$$d(1, m, M(1)) = M(1) ,$$

$$d(n+1, m, M(n+1)) = (g(N, m, M)+1)(M(1)+d(n, m, M(n)))$$

where $M = M(n+1)$ and $N = N(n+1, d(n, m, M(n)), M(1))$ as in (4), so N is the number of words in

$$Y_N = \{x_{n+1}^e v \mid 1 \le e \le M(1),\ v \in \mathrm{FM}(X_n),\ 1 \le \deg v \le d(n, m, M(n))\} .$$

The lemma will be proved by induction on n. It holds for $n = 1$ since in this case $d(1, m, M(1)) = M(1)$ and any word of degree greater than $d(1, m, M(1))$ contains the subword $x_1^{M(1)}$. Now assume the lemma for n and let $w \in \mathrm{FM}(X_{n+1})$ have degree greater than $d(n+1, m, M(n+1))$. Write

$$w = w_0 x_{n+1}^{e(1)} w_1 x_{n+1}^{e(2)} \dots x_{n+1}^{e(r)} w_r x_{n+1}^{e(r+1)}$$

where $e(i) > 0$ if $i < r+1$, $w_i \in \mathrm{FM}(X_n)$ and $\deg w_i > 0$ if $i > 0$. If some $e(i) > M(1)$ then w contains the subword $x_{n+1}^{M(1)}$ and we are done. If some w_i has degree greater than $d(n, m, M(n))$ we are done by the induction hypothesis. Thus we may assume every $e(i) \le M(1)$ and every $\deg w_i \le d(n, m, M(n))$. Then

$$\deg w \leq (r+1)\big(M(1)+d(n,\ m,\ M(n))\big)$$

and since

$$\deg w > d(n+1,\ m,\ M(n+1)) = \big(g(N,\ m,\ M)+1\big)\big(M(1)+d(n,\ m,\ M(n))\big)$$

it follows that

$$r > g(N,\ m,\ M)\ .$$

Hence we may assume that if Y_N is defined as above then w contains a Y_N-subword of Y-degree $r > g(N, m, M)$. Then the Second Combinatorial Lemma implies that w either contains a subword u^M , $u \in \mathrm{FM}(Y_N)$, where the Y-degree is positive and less than or equal to m or w has a subword that has a dominant factorization in $\mathrm{FM}(Y_N)$ (and hence in $\mathrm{FM}(X_{n+1})$) of length m . In the first case $\deg u \leq m\big(M(1) + d(n, m, M(n))\big) = \delta(n+1)$ and u is an $(m+1)$-word. In the second case we have a dominant factorization of length m into $(m+1)$-words. □

Theorem 2 can be extended also to algebras without unit. If A is an associative algebra without unit, a *special Jordan algebra without unit* J in A is a K-submodule of A such that if $a, b \in J$ then aba and $a^2 \in J$. Homomorphisms between such algebras are defined as K-module maps η such that $\eta(aba) = \eta(a)\eta(b)\eta(a)$ and $\eta(a^2) = \eta(a)^2$. Let $K\{X\}'$ denote the ideal in $K\{X\}$ generated by the x_i . This is the set of elements of $K\{X\}$ with zero constant term. Let $\mathrm{FSJ}(X)'$ denote the smallest K-submodule of $K\{X\}'$ containing X and closed under a^2 and aba . Then it is readily seen that if J is a special Jordan algebra without unit and $x_i \mapsto a_i$, $i = 1, 2, \ldots$, is a map of X into J then this has a unique extension to a homomorphism of $\mathrm{FSJ}(X)'$ into J . The various concepts we had before carry over. Moreover, the method used to prove Theorem 1 carries over to prove the following

THEOREM 3. *Let* J *be a special Jordan algebra without unit contained in an associative algebra* A *without unit such that* Env J = A *where* Env J *is the subalgebra of* A *generated by* J . *Assume* J *satisfies a monic identity and* J *has a set of generators* $\{a_i\}$ *such that every Jordan monomial in the* a_i *is nilpotent. Then* A *is locally nilpotent.*

The results can be applied also to algebraic and nil special Jordan algebras of bounded degree since it can be shown that these satisfy monic identities.

We shall conclude our discussion of the Jordan case by showing how the results apply to abstract Jordan algebras. For simplicity we restrict our attention to unital algebras. The concept of an (abstract) Jordan algebra arises in attempting to treat special ones in an intrinsic manner. This is highly desirable for many reasons. For one thing, the homomorphic image of a special Jordan algebra need not be special, so this class of algebras is not a variety. The definition of a (unital) Jordan algebra that is due to McCrimmon is as follows. A *Jordan algebra over* K is a triple $(J, U, 1)$ where J is a K-module, $1 \in J$ and U is a map of J into $\mathrm{End}_K J$ such that the following axioms hold:

QJ1. U is quadratic: $U_{\alpha a} = \alpha^2 U_a$ if $\alpha \in K$, $a \in J$.

QJ2. $U_1 = 1$.

QJ3. $U_a U_b U_a = U_{U_a b}$.

QJ4. If $U_{a,b} = U_{a+b} - U_a - U_b$ and $V_{a,b}$ is defined by $V_{a,b}x = U_{a,x}b$ then $U_a V_{b,a} = V_{a,b} U_a$.

QJ5. QJ3 and QJ4 hold "absolutely", that is, for all J_L where L is a commutative (associative) algebra over K.

If A is an associative algebra over K, A defines a Jordan algebra A^+ in which A is the K-module, the unit of A is the 1 in the definition and U_a is defined by $U_a x = axa$. If J and J' are Jordan algebras, a *homomorphism* η of J into J' is defined to be a K-module map such that $\eta(1) = 1'$, $\eta(U_a b) = U_{\eta(a)}\eta(b)$. The class of Jordan algebras over K with this definition of morphism constitutes a category. If A and A' are associative and η is a homomorphism of A into A' then η is a homomorphism of A^+ into A'^+. Then we have a functor from the category of associative algebras to the category of Jordan algebras. This functor has an adjoint. This fact amounts to the following: given any Jordan algebra J, there exists an associative algebra $s(J)$ and a homomorphism σ_u of J into $s(J)^+$ such that if η is any homomorphism of J into A^+ where A is associative then there exists a unique homomorphism ζ of associative algebras such that

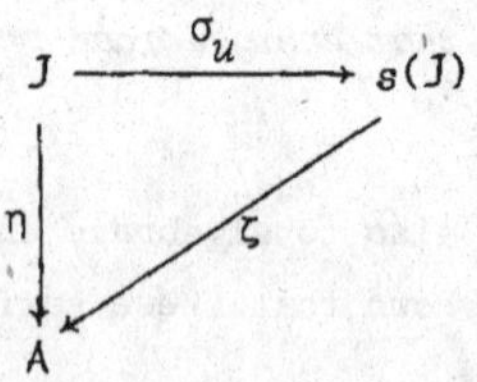

is commutative. $(s(J), \sigma_u)$ or simply $s(J)$ is called the *special universal algebra* of J. Evidently this is analogous to the universal enveloping algebra of a Lie algebra.

We shall obtain a consequence on $s(J)$ of Theorem 2. To formulate this we need to define powers, algebraic elements and monic identities for Jordan algebras. If $a \in J$ we define $a^0 = 1$, $a^1 = a$, $a^{n+2} = U_a a^n$. If $J = A^+$ then a^n defined in J coincides with a^n as defined in A. For arbitrary J we call an element $a \in J$ algebraic if there exists a monic polynomial $f(\lambda) \in K[\lambda]$ such that $f(a) = 0$ with the obvious meaning for $f(a)$. To define identities for Jordan algebras we need the concept of a free Jordan algebra $FJ(X)$ defined by $X = \{x_1, x_2, \ldots\}$. We shall not give a construction for this but will be content to invoke a general theorem on universal algebra that insures the existence of an $FJ(X)$ such that $FJ(X)$ contains X and any map of X into a Jordan algebra J has a unique extension to a homomorphism of $FJ(X)$ into J. In particular we have a unique homomorphism ν of $FJ(X)$ into $FSJ(X)$ such that $\nu x_i = x_i$, $i = 1, 2, 3, \ldots$. We call an element $f \in FJ(X)$ an *identity* for J if $f \mapsto 0$ under every homomorphism of $FJ(X)$ into J. It is known that $FSJ(X)$ has non-zero identities and these will be identities for all special Jordan algebras. We wish to avoid these and so we confine attention to identities f that are *monic* in the sense that νf is a monic element of $FSJ(X)$.

We can now obtain the abstract version of Theorem 2. We shall prove the following somewhat weaker result.

THEOREM 4. *Let J be a Jordan algebra such that:*

1. *J satisfies a monic identity, and*
2. *J is algebraic in the sense that every element of J is algebraic.*

Then the special universal envelope $s(J)$ is locally finite.

Proof. The image $\sigma_u(J)$ is a special Jordan algebra in $s(J)$ and it is well known that $\text{Env}\, \sigma_u(J) = s(J)$. Now we have the hypothesis of Theorem 2 for $\sigma_u(J)$ and $s(J)$. Hence the local finiteness of $s(J)$ follows from Theorem 2. □

References

Nathan Jacobson

[1] *Lectures on quadratic Jordan Algebras* (Tata Institute of Fundamental Research Lectures on Mathematics, **45**. Tata Institute of Fundamental Research, Bombay, 1969).

[2] *PI-Algebras. An Introduction* (Lecture Notes in Mathematics, **441**. Springer-Verlag, Berlin, Heidelberg, New York, 1975).

Kevin McCrimmon

[1] "Alternative algebras satisfying polynomial identities", *J. Algebra* 24 283-292.

А.И. Ширшов [A.I. Shirshov]

[1] "О некоторых неассоциативных нилькольцах и алгебраических алгебрах" [On some non-associative nil rings and algebraic algebras], *Mat. Sb. (N.S.)* 41 (83) (1957), 381-394.

[2] "О кольцах с тождественными соотношениями" [On rings with identical relations], *Mat. Sb. (N.S.)* 43 (85) (1957), 277-283.

M. Slater

[1] "Local functions for certain alternative algebras", Lecture Notes, University of Bristol, Bristol, 1967.

Department of Mathematics,
Yale University,
New Haven,
Connecticut,
USA.

PROC. 18th SRI
CANBERRA 1978, 47-51.

20-02
(20K99)

FIVE THEOREMS ON ABELIAN GROUPS

Irving Kaplansky

1. Introduction

In 1961, Dartmouth hosted a conference entitled "New Directions in Mathematics" and in due course the proceedings were published [142]. My talk contained some predictions. Two seem to be working out well: I predicted that by 1984 there would be substantial progress on finite simple groups and on Albert's problems concerning associative division algebras. (Why 1984? The speakers had been invited to look ahead a quarter of a century and I rounded off 25 to the nearest prime.)

On pages 108-109 I made some remarks about infinite abelian groups and offered the suggestion that the subject might be allowed to take a rest for a while. I was not serious. In fact, seven years later I prepared a second edition of my *Infinite Abelian Groups*. Moreover, my advice was fortunately not taken. Of the many fine achievements in abelian group theory, I have in this article picked five theorems I particularly admire which were proved since my 1961 indiscretion.

My account will be short and will not carry the reader to the frontiers of the subject. A vivid picture of today's activity in the field of abelian groups is provided by the papers in [127]. (The numbering starts at [127] since I shall also be referring to the bibliography of the 1968 edition of *Infinite Abelian Groups*.)

This written version is reasonably close to the oral presentation I made to the 1978 Summer Research Institute of the Australian Mathematical Society which, in turn, was a variant of talks presented at Santa Barbara in April 1976 and at Saint Andrews in July, 1976. I take the opportunity to thank everyone involved (on all three occasions) for hospitality and superb arrangements.

2. The stacked bases theorem

In proving the basis theorem for a finitely generated abelian group A one can operate as follows: write $A = F/G$ with F free (and therefore G also free) and show that "stacked bases" can be chosen for F and G. By this one means that F has a basis $u_1, \ldots, u_n$ such that the elements $a_1u_1, \ldots, a_nu_n$ form a basis of G for suitable integers $a_1, \ldots, a_n$.

Now drop the assumption that A is finitely generated. Of course, to have any hope of writing $A = F/G$ with stacked bases one must assume outright that A is a direct sum of cyclic groups. Can it then be done? I asked this question in connection with Exercise 80 of *Infinite Abelian Groups* and it survived intact through the second edition. Cohen and Gluck [130] proved that the answer is "yes". I have no further comment except to point out that it took two topologists to do it.

The remaining four topics share a familiar feature of abelian group theory: they are of interest only in the uncountable case.

3. Totally projective groups

This class of primary abelian groups has at least three different basic definitions, and it is non-trivial to prove that the various versions are equivalent. So I shall take the easy way out and give no definition. At any rate, Ulm's theorem generalizes to totally projective groups; this means that a certain function from ordinals to cardinals classifies the groups completely. (Ulm's theorem was the countable case.) In a sense [135, Theorem 70] totally projective groups constitute a final word on this subject.

It is regretable that Hill's important work is not available under his own name. The paper [53] and the paper quoted in the bibliographies of [134] and [135] under the different title "On the classification of abelian groups" have not appeared. However, it is believed that [135] gives a tolerably faithful account. By consulting it, the paper [144], and Chapter 12 in [134], a reader can become reasonably well informed concerning the subject.

Many of the papers in [127] report on recent extensions of the theory to non-torsion abelian groups.

4. Baer's problem

Let G be a torsion-free group such that $\mathrm{ext}(G, T) = 0$ for every torsion group T. In other words, we are requiring that a mixed abelian group must split whenever

G is the torsion-free factor group modulo the torsion subgroup. This is certainly true if G is free, for then we get splitting when G is combined with any group, torsion or not. Does the hypothesis force G to be free? The answer is "yes" and Griffith [47] did it.

It is interesting to note that this problem asks whether a certain hypothesis forces an abelian group to be free, and the next two do precisely the same thing. Hill's paper in [127] has some cogent comments on this situation.

5. Specker's problem

Let G be the group of all bounded sequences of integers, under pointwise addition. (It is perhaps more revealing to think of G as all finite-valued functions from a countably infinite set to the integers.) In [117] Specker asked whether G is free and proved that every subgroup of cardinal $\aleph_1$ is free. (Yes, I mean $\aleph_1$.) Since the cardinal number of G is the continuum, if we are willing to assume the continuum hypothesis we can wrap the problem up and go home.

More cautiously, one said that Specker had left the problem in the following state: either the answer was affirmative or the question was undecidable. Several times I brazenly expressed the opinion that the question was too "down to earth" to be undecidable and so somebody, some day would prove it.

REMARKS 1. I don't think I ever said anything similar about Whitehead's problem.

2. See [131] for a vaguely related letter.

Well, anyway, Nöbeling [141] proved it and Bergman [129] and Hill [136] simplified the proof and generalized the theorem. See also [137] for an application.

6. Whitehead's problem

The statement is deceptively similar to that of Baer's problem: does $\mathrm{ext}(G, \mathbb{Z}) = 0$ force G to be free? This time we are requiring splitting when the subgroup is infinite cyclic, and the factor group is G. The answer is "yes" if G is countable. Better still, every countable subgroup of G is free, and this is pretty easy to prove. (For many years Baer's problem was in the same state.)

Well, what's the answer? Hold on to your hats, fellow non-logicians: Shelah [143] has proved this to be undecidable. If one version of set theory is assumed, the answer is "yes"; if another, it is "no".

In a little greater detail: there is a certain statement about partially ordered sets called "Martin's axiom". If Martin's axiom is assumed and the continuum hypothesis is denied then there is a non-free group of cardinal $\aleph_1$ satisfying the

Whitehead hypothesis. On the other hand, there is a certain statement about ordinals which it would seem reasonable to call "Jensen's axiom". Jensen's axiom implies that every group of cardinal $\aleph_1$ satisfying the hypothesis is free. (I am essentially quoting the excellent article of Eklof [132]. See Nunke's article in [127] for further background and history.)

The assertions made in the preceding paragraph are conventional theorems. The metamathematics comes in to assure us that the set-theoretic assumptions in question are consistent with the usual axioms of set theory. I should add that Jensen proved that this "axiom" is a consequence of $V = L$, this being the assumption that all sets are constructible, used by Gödel in proving the consistency of the continuum hypothesis.

Bibliography

In addition to items to which there is a reference, I have inserted the general references 128, 138, 139, and 140.

[127] D. Arnold, R. Hunter, and E. Walker [Eds.], *Abelian Group Theory* (Proc. 2nd New Mexico State University Conference, Las Cruces, New Mexico, 1976. Lecture Notes in Mathematics, **616**. Springer-Verlag, Berlin, Heidelberg, New York, 1977).

[128] Gilbert Baumslag [Ed.], *Reviews on Infinite Groups*, 2 (Amer. Math. Soc., Providence, Rhode Island, 1974). (Section XVIII, pages 616-736 is devoted to abelian groups. MR reviews from 1940 through 1970, vols. 1-40 are included.)

[129] George M. Bergman, "Boolean rings of projection maps", *J. London Math. Soc.* (2) **4** (1972), 593-598.

[130] Joel M. Cohen and Herman Gluck, "Stacked bases for modules over principal ideal domains", *J. Algebra* **14** (1970), 493-505.

[131] Richard M. Dudley, "Unsolvable problems in mathematics" (Letter to the Editor), *Science* **191** (1976), 807-808.

[132] Paul C. Eklof, "Whitehead's problem is undecidable", *Amer. Math. Monthly* **83** (1976), 775-788.

[133] László Fuchs, *Infinite Abelian Groups*, I (Pure and Applied Mathematics, **36**. Academic Press, New York, London, 1970).

[134] László Fuchs, *Infinite Abelian Groups*, II (Pure and Applied Mathematics, **36-II**. Academic Press, New York, London, 1973).

[135] Phillip A. Griffith, *Infinite Abelian Group Theory* (University of Chicago Press, Chicago, Illinois; London; 1970).

[136] Paul Hill, "The additive group of commutative rings generated by idempotents", *Proc. Amer. Math. Soc.* **38** (1973), 499-502.

[137] P. Hill and H. Subramanian, "The freeness of a group based on a distributive lattice", *Proc. Amer. Math. Soc.* **51** (1975), 260-262.

[138] А.Г. Курош [A.G. Kurosh], Теория Групп [*The Theory of Groups*], 3rd edition (Izdat. "Nauka", Moscow, 1967). (This edition has expanded coverage on abelian groups: pages 144-203 and 549-578.)

[139] A.P. Mišina, "Abelian groups", *Progress in Mathematics*, 5, 1-37 (Plenum Press, New York, London, 1969).

[140] A.P. Mishina, "Abelian groups", *J. Soviet Math.* **2** (1974), 239-263.

[141] G. Nöbeling, "Verallgemeinerung eines Satzes von Herrn E. Specker", *Invent. Math.* **6** (1968/69), 41-55.

[142] Robert W. Ritchie [Ed.], *New Directions in Mathematics* (Dartmouth College Mathematics Conference, 1961. Prentice Hall, Englewood Cliffs, New Jersey, 1963).

[143] Saharon Shelah, "Infinite abelian groups, Whitehead problem and some constructions", *Israel J. Math.* **18** (1974), 243-256.

[144] Elbert A. Walker, "Ulm's theorem for totally projective groups", *Proc. Amer. Math. Soc.* **37** (1973), 387-392.

Note added in proof (5 June, 1978). 1. For an application of Nöbeling's theorem to topology see W.S. Massey, "How to give an exposition of the Čech-Alexander-Spanier type homology theory", *Amer. Math. Monthly* **85** (1978), 75-83. 2. A recent additional reference concerning the Whitehead problem is Saharon Shelah, "Whitehead groups may not be free, even assuming *CH*, I", *Israel J. Math.* **28** (1977), 193-204.

Department of Mathematics,
University of Chicago,
Chicago,
Illinois,
USA.

PROC. 18th SRI
CANBERRA 1978, 52-69.

20C10

INTEGRAL REPRESENTATIONS:
GENUS, *K*-THEORY AND CLASS GROUPS

Irving Reiner

1. Introduction

Let R be a ring of integers with quotient field K of characteristic 0. We assume throughout that R is a Dedekind ring, such as Z, or more generally the ring alg.int.$\{K\}$ of all algebraic integers in an algebraic number field K. Such rings are called *global* rings of integers. We will also consider the cases where R is a *local* ring, that is, a localization or completion of a global ring.

Given a finite group G, we want to consider representations of G by matrices with entries in R. Equivalently, we may study modules over the integral group ring RG; those modules with a free finite R-basis yield matrix representations. However, we need a more general framework in which to consider such problems. Instead of RG we shall deal with an R-order Λ in a K-algebra A, and instead of R-free RG-modules we shall consider Λ-lattices. We recall some definitions:

Let A be a finite dimensional K-algebra; an *R-order* in A is a subring Λ of A such that

(i) $R \subseteq$ center of Λ,

(ii) Λ is f.g./R (finitely generated as R-module), and

(iii) $K\cdot\Lambda = A$ (that is, Λ spans A over K).

EXAMPLES. (a) $A = KG$, $\Lambda = RG$, where G is any finite group.

(b) If K is an algebraic number field, then alg.int.$\{K\}$ is a Z-order in the

Q-algebra K.

(c) $A = M_n(K)$, $\Lambda = M_n(R)$, where $M_n(\ )$ means ring of $n \times n$ matrices.

Let Λ be an R-order in the K-algebra A. A *Λ-lattice* is a left Λ-module which is f.g./R and R-torsionfree (or equivalently, f.g./R and R-projective). Each Λ-lattice M may be embedded in the A-module $K \otimes_R M$, which we then write as KM. Thus, Λ-lattices are just finitely generated Λ-submodules of A-modules. The fundamental problem in integral representation theory is to classify all Λ-lattices for a given order Λ. Such information can then be used to solve questions in group theory, algebraic number theory, and algebraic topology.

One method of studying Λ-lattices is to reduce the global problem to local problems; this is a standard technique in algebra. For P a maximal ideal of the Dedekind ring R, we may define a P-adic valuation on the field K, and then form completions with respect to the non-archimedean metric associated with the P-adic valuation. Let K_P denote the P-adic completion of K, and R_P the P-adic completion of R; then R_P is a complete discrete valuation ring with quotient field K_P.

For each P, we may form

$$A_P = K_P \otimes_K A, \quad \Lambda_P = R_P \otimes_R \Lambda .$$

Then Λ_P is an R_P-order in the K_P-algebra A_P. For a Λ-lattice M, set

$$M_P = R_P \otimes_R M = \Lambda_P\text{-lattice}.$$

DEFINITION. Let M, N be Λ-lattices. Then M and N are in the same *genus* (notation: $M \vee N$) if for each maximal ideal P of R, there is a Λ_P-isomorphism $M_P \cong N_P$. (We also say that M is *locally isomorphic* to N.)

Instead of using the completion M_P, one could also work with the localization $(R-P)^{-1}M$ of M relative to the multiplicative subset $R - P$ of R. However, for each P we have

$$M_P \cong N_P \Longleftrightarrow (R-P)^{-1}M \cong (R-P)^{-1}N,$$

so genus can equally well be defined via localization. Nevertheless, there are significant advantages in using P-adic completions. For example, the technique of "lifting idempotents" works P-adically; as a consequence of this fact, one obtains

THEOREM (Krull-Schmidt-Azumaya). *Every finitely generated Λ_P-module is expressible as a finite direct sum of indecomposable modules, with the summands unique*

up to isomorphism and order of occurrence.

In order to decide whether two Λ-lattices M and N are in the same genus, we need to look at their completions at all primes P. It turns out that when A is a semisimple K-algebra, there is some critical finite set S of primes, such that $M \vee N$ if and only if $M_P \cong N_P$ for each $P \in S$. For example, when $A = KG$ we may choose S to be the set of maximal ideals P of R which contain $|G|$. The general case can be handled most easily by using maximal orders and their properties, which we now review briefly.

Let Λ be an R-order in the K-algebra A; call Λ a *maximal* R-order if there is no larger R-order in A which properly contains Λ. Under the hypothesis that A is semisimple, it is known that maximal orders exist; indeed, given *any* R-order Λ_0 in A, there exists at least one maximal order Λ in A containing Λ_0. (If char $K \neq 0$, the hypothesis that A be semisimple must be replaced by the stronger hypothesis that A is a separable K-algebra, that is, $E \otimes_K A$ is semisimple for each extension field E of K.)

Let Λ be a maximal R-order in a semisimple K-algebra A. The following facts are not difficult to establish:

(i) for each P, the completion Λ_P is a maximal R_P-order in A_P;

(ii) every one-sided ideal of Λ is projective as Λ-module;

(iii) every Λ-lattice is projective as Λ-module, and is isomorphic to a finite external direct sum of left ideals of Λ;

(iv) let X, Y be Λ_P-lattices: then $X \cong Y$ if and only if $K_P X \cong K_P Y$.

Now let Λ be an arbitrary R-order in the semisimple K-algebra A, and let $\Lambda \subseteq \Lambda' \subset A$, where Λ' is a maximal R-order. We can choose a nonzero element $\alpha \in R$ such that $\alpha\Lambda' \subseteq \Lambda \subseteq \Lambda'$. Then for each maximal ideal P of R which does not contain α, the element α is a unit in the completion R_P, and therefore $\Lambda'_P = \Lambda_P$. It follows at once that $\Lambda'_P = \Lambda_P$ almost everywhere (that is, for all but a finite number of P's). Let

$$S = \{P : P = \text{maximal ideal of } R,\ \Lambda_P \neq \text{maximal } R_P\text{-order in } A_P\}.$$

Then S is a finite set.

LEMMA. *Let M, N be Λ-lattices such that $M_P \cong N_P$ for each $P \in S$, and such that $KM \cong KN$. Then $M \vee N$.*

Proof. For $P \notin S$, Λ_P is a maximal order. Since $KM \cong KN$, it follows that $K_P M \cong K_P N$. But

$$K_P M = K_P \otimes_K (K \otimes_R M) \cong K_P \otimes_{R_P} (R_P \otimes_R M) = K_P M_P ,$$

so $K_P M_P \cong K_P N_P$. Therefore $M_P \cong N_P$ by (iv) above. This shows that $M_P \cong N_P$ for every $P \notin S$. On the other hand, $M_P \cong N_P$ for $P \in S$ by hypothesis, and so $M \vee N$. We may remark that if S is non-empty, then the hypothesis that $KM \cong KN$ can be omitted. Indeed, if $M_P \cong N_P$ for some specific P , then we obtain $K_P M \cong K_P N$, that is,

$$K_P \otimes_K (KM) \cong K_P \otimes_K (KN) .$$

It follows that $KM \cong KN$ by the Noether-Deuring Theorem.

We remark that when $\Lambda = RG$, an integral group ring, then Λ_P is a maximal order if and only if P does not contain $|G|$. Hence in this case, the exceptional set S consists of all prime ideal divisors of $|G|$.

The following criterion is useful in discussing genera:

ROITER'S LEMMA. *Let* M, N *be* Λ*-lattices. Then* $M \vee N$ *if and only if for each finite set* S_0 *of prime ideals of* R *, there exists a short exact sequence of* Λ*-modules*

$$0 \to M \to N \to T \to 0$$

such that $T_P = 0$ *for each* $P \in S_0$.

As an application of this result (whose proof we omit), we show how to "add" lattices in the same genus.

PROPOSITION. *Let* L, M, N *be* Λ*-lattices in the same genus, where* Λ *is any* R*-order. Then there exists a* Λ*-lattice* L' *in the genus, such that*

$$M \oplus N \cong L \oplus L' .$$

Proof. By Roiter's Lemma, we can find a pair of Λ-exact sequences

$$0 \to M \xrightarrow{f} L \to T \to 0 , \quad 0 \to N \xrightarrow{g} L \to U \to 0 ,$$

such that T, U are R-torsion Λ-modules, and such that for each maximal ideal P of R , either $T_P = 0$ or $U_P = 0$. Now consider the exact sequence

$$0 \to L' \to M \oplus N \xrightarrow{(f,g)} L$$

where $L' = \ker(f, g)$. Then for each P , either the map $f_P : M_P \to L_P$ is surjective or the map $g_P : N_P \to L_P$ is surjective. Hence $(f, g)_P$ is surjective for each P , whence so is the map (f, g) . We thus obtain a short exact sequence

$$0 \to L' \to M \oplus N \to L \to 0$$

which splits at each P, and hence splits globally. This establishes the proposition.

It is instructive to compare the preceding result with Steinitz's Theorem on the structure of R-lattices, where R is any Dedekind domain. An *R-lattice* is a finitely generated torsionfree R-module. If $\underline{\underline{a}}$ and $\underline{\underline{b}}$ are nonzero ideals of R, we may form their product $\underline{\underline{ab}}$, a nonzero ideal consisting of all finite sums $\sum a_i b_i$, $a_i \in \underline{\underline{a}}$, $b_i \in \underline{\underline{b}}$. The ideals $\underline{\underline{a}}, \underline{\underline{b}}$ are said to be in the same *ideal class* if $\underline{\underline{a}} = \underline{\underline{b}}x$ for some nonzero $x \in K$, where K is the quotient field of R. (We note that $\underline{\underline{a}}$ and $\underline{\underline{b}}$ are in the same ideal class if and only if $\underline{\underline{a}} \cong \underline{\underline{b}}$ as R-modules.) The *rank* of an R-lattice M is defined to be $\dim_K(K \otimes_R M)$.

STEINITZ'S THEOREM. *Let M be an R-lattice of rank n. Then there exist nonzero ideals $\{\underline{\underline{a}}_i : 1 \le i \le n\}$ of R, such that*

$$M \cong \underline{\underline{a}}_1 \oplus \dots \oplus \underline{\underline{a}}_n \quad \textit{(external direct sum).}$$

Furthermore,

$$\underline{\underline{a}}_1 \oplus \dots \oplus \underline{\underline{a}}_n \cong \underline{\underline{b}}_1 \oplus \dots \oplus \underline{\underline{b}}_m$$

if and only if $m = n$ and the products $\underline{\underline{a}}_1 \dots \underline{\underline{a}}_n$ and $\underline{\underline{b}}_1 \dots \underline{\underline{b}}_m$ lie in the same ideal class.

Now for each maximal ideal P of R, the completion R_P is a principal ideal ring. Hence for each nonzero ideal $\underline{\underline{a}}$ of R, we have $\underline{\underline{a}}_P \cong R_P$ as R_P-modules. In particular, let M be an R-lattice of rank n. Then for each P we have

$$M_P \cong R_P^{(n)} \quad \text{(direct sum of } n \text{ copies of } R_P\text{)}.$$

This shows that $M \vee R^{(n)}$, so the genus of an R-lattice M depends only on its rank. If $\underline{\underline{a}}$ and $\underline{\underline{b}}$ are a pair of nonzero ideals of R, then each has rank 1, so $\underline{\underline{a}}, \underline{\underline{b}}$ and R lie in the same genus. It follows from the proposition preceding Steinitz's Theorem that

$$\underline{\underline{a}} \oplus \underline{\underline{b}} \cong R \oplus \underline{\underline{c}} \quad \text{for some nonzero ideal } \underline{\underline{c}} \text{ of } R.$$

Steinitz's Theorem gives a more precise form of this assertion, namely:

$$\underline{\underline{a}} \oplus \underline{\underline{b}} \cong R \oplus \underline{\underline{ab}}\,.$$

Returning to the general situation, we point out next the fact that decomposability of a Λ-lattice depends only on its genus. Specifically, we have

PROPOSITION. *Let L, M, N be Λ-lattices such that $L \vee (M \oplus N)$. Then there exist Λ-lattices X, Y such that*

$$L \cong X \oplus Y , \quad X \vee M , \quad Y \vee N .$$

There is a much deeper result due to Roiter and Jacobinski, as follows:

THEOREM. *Let* Λ *be an* R*-order in a semisimple* K*-algebra* A *, where* K *is an algebraic number field. Let* F *be any faithful* Λ*-lattice, that is, no nonzero element of* Λ *annihilates* F *. Then for any pair of* Λ*-lattices* M, M' *in the same genus,* M' *is a direct summand of* $M \oplus F$ *. Indeed,*

$$M \oplus F \cong M' \oplus F' \quad \textit{for some} \quad F' \vee F .$$

REMARK. The hard part is to prove that M' is a direct summand of $M \oplus F$. Once this is known, the isomorphism $M \oplus F \cong M' \oplus F'$ easily implies that $F' \vee F$.

To conclude this introduction, we recall the connection between lattices and their completions:

PROPOSITION. *Let* M *be a* Λ*-lattice, where* Λ *is an* R*-order in a finite dimensional* K*-algebra* A *. Then*

$$M = KM \cap \left\{ \bigcap_P M_P \right\} ,$$

where P *ranges over all maximal ideals of* R *.*

Conversely, we wish to construct a Λ-lattice with given completions. The basic result of this type is as follows:

PROPOSITION. *Let* V *be a finitely generated* A*-module, and let* M *be a* Λ*-lattice in* V *such that* $KM = V$ *. Suppose that for each* P *we are given a* Λ_P*-lattice* $X(P)$ *in* V_P *such that*

(i) $K_P \cdot X(P) = V_P$ *for all* P *, and*

(ii) $X(P) = M_P$ *almost everywhere.*

Set

$$L = V \cap \left\{ \bigcap_P X(P) \right\} .$$

Then L *is a* Λ*-lattice in* V *such that*

$$K{\cdot}L = V , \quad L_P = X(P) \quad \textit{for all} \quad P .$$

The proof is not difficult, and is essentially a generalization of the "strong approximation theorem" of algebraic number theory.

2. Class groups

As before, we start with a Dedekind ring R with quotient field K of characteristic 0 , and now let Γ be an R-order in a finite dimensional K-algebra

B . Given a left Γ-lattice M , we wish to consider Γ-lattices N in the genus of M . We have remarked above that if $M_P \cong N_P$ for some specific maximal ideal P of R , then necessarily $KM \cong KN$. Replacing N by an isomorphic copy if need be, we may then assume that $N \subseteq KM$. Thus, in order to calculate all isomorphism classes of Γ-lattices N in the genus of M , we need only consider those lattices in the genus of M which are contained in KM . Of course, not every lattice in KM need lie in the same genus as M .

We now set $\Lambda = \text{end}_\Gamma(M)$, the ring of Γ-endomorphisms of M , which we view as right operator domain on M . Thus M becomes a left Γ-, right Λ-bimodule, and Λ is an R-order in the K-algebra A , where $A = \text{end}_B(KM)$. Let $u(A)$ denote the group of units of A . If N is a Γ-lattice in KM , then any Γ-map $\varphi : N \to M$ extends to a B-homomorphism $KN \to KM$, and hence φ is given by right multiplication by some element of A . Thus $N \cong M$ as Γ-lattices if and only if $N = Mx$ for some $x \in u(A)$.

Now let N be a Γ-lattice, $N \vee M$, $N \subseteq KM$. Then for each maximal ideal P of R we have $N_P \subseteq K_P M_P$, $N_P \cong M_P$, so by the preceding discussion we have

$$N_P = M_P\alpha_P \quad \text{for some} \quad \alpha_P \in u(A_P) .$$

Furthermore, there exist nonzero elements $r, s \in R$ such that $rM \subseteq N \subseteq sM$, which implies that $N_P = M_P$ almost everywhere. Hence we may choose $\alpha_P = 1$ almost everywhere, or somewhat less restrictively, we may assume that $\alpha_P \in u(\Lambda_P)$ almost everywhere. We have (forming intersections within KM)

$$N = \bigcap_P N_P = \bigcap_P M_P\alpha_P .$$

This suggests that we introduce the idele group $J(A)$ relative to R , defined as

$$J(A) = \left\{(\alpha_P) \in \prod_P u(A_P) \; : \; \alpha_P \in u(\Lambda_P) \text{ almost everywhere}\right\} .$$

(This definition is independent of the choice of the R-order Λ in A .) For $\alpha = (\alpha_P) \in J(A)$, define

$$\Lambda\alpha = \bigcap_P \Lambda_P\alpha_P \quad (\text{intersection within } A).$$

By the remarks at the end of §1, we know that $\Lambda\alpha$ is a left Λ-lattice in A such that

$$K\cdot\Lambda\alpha = A \; , \quad (\Lambda\alpha)_P = \Lambda_P\alpha_P \quad \text{for each } P .$$

Thus $\Lambda\alpha$ is isomorphic to a left ideal of Λ , and $\Lambda\alpha$ is in the same genus as Λ .

We call $\Lambda\alpha$ a *locally free* rank 1 Λ-lattice.

We may form $M\cdot\Lambda\alpha$, calculated within KM. We have (for each P)

$$(M\cdot\Lambda\alpha)_P = M_P\cdot(\Lambda\alpha)_P = M_P\cdot\Lambda_P\alpha_P = M_P\alpha_P = N_P ,$$

so $M\cdot\Lambda\alpha$ and N are a pair of left Γ-lattices in KM whose completions are the same for each P. It follows at once that

$$N = M\cdot\Lambda\alpha .$$

Thus, the Γ-lattices $M\cdot\Lambda\alpha$ represent all isomorphism classes of Γ-lattices in the genus of M. It is an easy task to decide when $M\cdot\Lambda\alpha \cong M\cdot\Lambda\beta$ as Γ-lattices, where $\alpha, \beta \in J(A)$. Indeed, if such an isomorphism holds true, then

$$M\cdot\Lambda\beta = (M\cdot\Lambda\alpha)a \quad \text{for some} \quad a \in u(A) .$$

This implies that for each P, there exists a unit $u_P \in u(\Lambda_P)$ such that

$$\beta_P = u_P\alpha_P a \quad \text{in} \quad A_P .$$

Conversely, such an equality (for each P) implies that $M\cdot\Lambda\alpha \cong M\cdot\Lambda\beta$.

The image of $u(A)$ in $J(A)$ is called the group of *principal idèles*, and is again denoted by $u(A)$. On the other hand, the group of *unit idèles* is defined as

$$U(\Lambda) = \prod_P u(\Lambda_P) .$$

We have thus established

PROPOSITION. *Let* M *be a left* Γ*-lattice, and let* $\Lambda = \text{end}_\Gamma(M)$, $A = K\Lambda$. *Then there is a bijection between the set of isomorphism classes of* Γ*-lattices in the genus of* M*, and the set of double cosets* $U(\Lambda)\cdot\alpha\cdot u(A)$ *of the idèle group* $J(A)$. *The double coset* $U(\Lambda)\cdot\alpha\cdot u(A)$ *corresponds to the isomorphism class of the* Γ*-lattice*

$$M\cdot\Lambda\alpha = KM \cap \left\{\bigcap_P M_P\alpha_P\right\} .$$

We have now shown that the study of Γ-lattices in the genus of a given left Γ-lattice M is equivalent to the study of left Λ-lattices in the genus of Λ, where $\Lambda = \text{end}_\Gamma(M)$. Each left Λ-lattice in the genus of Λ is (up to isomorphism) of the form $\Lambda\alpha$, where $\alpha \in J(A)$ is an idèle, and where

$$\Lambda\alpha = A \cap \left\{\bigcap_P \Lambda_P\alpha_P\right\} , \quad \alpha = (\alpha_P) .$$

Further, if $\alpha, \beta \in J(A)$, then

$$\Lambda\alpha \cong \Lambda\beta \quad \text{if and only if} \quad \beta \in U(\Lambda)\cdot\alpha\cdot u(A) .$$

Now let $\alpha, \beta \in J(A)$; since $\Lambda\alpha, \Lambda\beta$ and Λ are Λ-lattices in the same genus, there must exist an idèle $\gamma \in J(A)$ for which

$$\Lambda\alpha \oplus \Lambda\beta \cong \Lambda \oplus \Lambda\gamma .$$

In fact, we prove

PROPOSITION. *For* $\alpha, \beta \in J(A)$, *we have*

$$\Lambda\alpha \oplus \Lambda\beta \cong \Lambda \oplus \Lambda\alpha\beta$$

Proof. Replacing α by $r\alpha$ with nonzero $r \in R$, we may assume α integral, that is, $\alpha_P \in \Lambda_P$ for each P . (Such a change does not affect the isomorphism classes of $\Lambda\alpha$ or $\Lambda\alpha\beta$.) Further, we may replace α by $u\alpha$, $u \in U(\Lambda)$, without affecting these isomorphism classes; thus we may assume that $\alpha_P = 1$ almost everywhere from now on. Let

$$S = \{P : \alpha_P \neq 1\} ,$$

a finite set. Choose a positive integer k such that

$$\alpha_P^{\pm 1} \in P^{-k}\Lambda_P , \quad P \in S .$$

Next, we may replace β by βx , $x \in u(A)$, without affecting the isomorphism classes of $\Lambda\beta$ or $\Lambda\alpha\beta$. Since $u(A)$ is dense in $J(A)$, we may choose $x \in u(A)$ so that

$$\beta_P x \equiv 1 \left(\bmod P^t\Lambda_P\right) , \quad P \in S ,$$

where $t = 2k + 1$. Replacing β by βx and changing notation, we may assume that β is an integral idèle such that

$$\beta_P \equiv 1 \left(\bmod P^t\Lambda_P\right) , \quad P \in S .$$

Since $P\Lambda_P \subseteq \operatorname{rad} \Lambda_P$, it follows that $\beta_P \in u(\Lambda_P)$, $P \in S$. Further, for $P \in S$, we have

$$\beta_P\alpha_P\beta_P^{-1}\alpha_P^{-1} = \left(1+\lambda\pi^t\right)\alpha_P \cdot \left(1+\lambda_1\pi^t\right)\alpha_P^{-1} , \quad \lambda, \lambda_1 \in \Lambda_P ,$$

where $PR_P = \pi R_P$. Since $\alpha_P^{\pm 1} \in P^{-k}\Lambda_P$, a simple calculation then shows that

$$\beta_P\alpha_P\beta_P^{-1}\alpha_P^{-1} = 1 + \lambda_2\pi^{t-2k} , \quad \lambda_2 \in \Lambda_P ,$$

whence the commutator is a unit in Λ_P . This gives

$$\Lambda_P\alpha_P\beta_P = \Lambda_P\beta_P\alpha_P , \quad P \in S .$$

Now we claim that $\Lambda\alpha + \Lambda\beta = \Lambda$; it suffices to check that $\Lambda_P\alpha_P + \Lambda_P\beta_P = \Lambda_P$.

But this is clear since $\alpha_P, \beta_P \in \Lambda_P$ for all P, and for each P either α_P or $\beta_P \in u(\Lambda_P)$. We thus obtain a Λ-exact sequence

$$0 \to L \to \Lambda\alpha \oplus \Lambda\beta \xrightarrow{g} \Lambda \to 0,$$

where g is induced by the inclusion maps $\Lambda\alpha \to \Lambda$, $\Lambda\beta \to \Lambda$, and $L = \ker g$. We have

$$L \cong \bigcap_P (\Lambda_P\alpha_P \cap \Lambda_P\beta_P).$$

For $P \in S$ we have $\beta_P \in u(\Lambda_P)$, so

$$\Lambda_P\alpha_P \cap \Lambda_P\beta_P = \Lambda_P\alpha_P = \Lambda_P\beta_P\alpha_P = \Lambda_P\alpha_P\beta_P.$$

For $P \notin S$ we have $\alpha_P = 1$, so

$$\Lambda_P\alpha_P \cap \Lambda_P\beta_P = \Lambda_P\beta_P = \Lambda_P\alpha_P\beta_P.$$

Hence $L \cong \Lambda\alpha\beta$. The preceding exact sequence splits since Λ is projective, and so we obtain

$$\Lambda\alpha \oplus \Lambda\beta \cong \Lambda \oplus \Lambda\alpha\beta$$

as desired.

This result was originally proved by Fröhlich using reduced norms (see [5] for references). One surprising consequence is that

$$\Lambda \oplus \Lambda\beta\alpha \cong \Lambda \oplus \Lambda\alpha\beta \quad \text{for all } \alpha, \beta \in J(A).$$

Since $(\Lambda\alpha)_P \cong \Lambda_P$ for each P, we call $\Lambda\alpha$ a *locally free* Λ-lattice (of rank 1). By analogy with the ideal class group of a Dedekind ring, we wish to introduce the locally free class group of Λ, denoted by $\mathrm{cl}(\Lambda)$. The elements of $\mathrm{cl}(\Lambda)$ should be "classes" of locally free Λ-lattices of rank 1, with "addition" of classes given by the formula

$$[\Lambda\alpha] + [\Lambda\beta] = [\Lambda\alpha\beta].$$

However, the isomorphism classes of $\Lambda\alpha$ and $\Lambda\beta$ do not (in general) determine the isomorphism class of $\Lambda\alpha\beta$, since it may happen that

$$\Lambda \oplus \Lambda\alpha\beta \cong \Lambda \oplus \Lambda\gamma, \text{ but } \Lambda\alpha\beta \ncong \Lambda\gamma.$$

To overcome this difficulty, we introduce the notion of stable isomorphism: two projective Λ-lattices M, N are called *stably isomorphic* if

$$M \oplus \Lambda^{(k)} \cong N \oplus \Lambda^{(k)} \quad \text{for some } k.$$

(In fact, such an isomorphism holds if and only if $M \oplus \Lambda \cong N \oplus \Lambda$.)

Suppose from now on that K is an algebraic number field, and that Λ is an

R-order in a semisimple K-algebra A. Jacobinski has shown that for M, N locally free Λ-lattices, stable isomorphism implies isomorphism, provided that no simple component of A is a totally definite quaternion algebra. In any case, let $[\Lambda\alpha]$ denote the stable isomorphism class of the locally free rank 1 Λ-lattice $\Lambda\alpha$, where $\alpha \in J(A)$. The set of such classes then forms an abelian additive group, with identity element $[\Lambda]$, and with $-[\Lambda\alpha] = [\Lambda\alpha^{-1}]$, $\alpha \in J(A)$. This group $\mathrm{cl}(\Lambda)$, the *locally free ideal class group* of Λ, can be expressed as a quotient of $J(A)$. (It follows trivially from the Jordan-Zassenhaus Theorem that $\mathrm{cl}(\Lambda)$ is finite.)

THEOREM (Jacobinski, Fröhlich). *Let $J'(A)$ denote the topological closure in the idèle group $J(A)$ of the commutator subgroup of $J(A)$. Then*

$$U(\Lambda)\cdot J'(A)\cdot u(A) \trianglelefteq J(A),$$

and

$$J(A)/U(\Lambda)\cdot J'(A)\cdot u(A) \cong \mathrm{cl}(\Lambda).$$

The isomorphism maps the coset containing the idèle α onto the stable isomorphism class of the locally free rank 1 Λ-lattice $\Lambda\alpha$.

The result was first proved by Jacobinski in a somewhat different form, and then more recently by Fröhlich. Both proofs depend strongly on the use of reduced norms and Eichler's Theorem. In practice, however, a slightly weaker version of this formula is more useful, and can be derived without appealing to such deep results from class field theory. This calculation is due to Wilson, and is given in [5], where a complete list of references can be found. Here we shall content ourselves with giving the final version of the formula for the class group $\mathrm{cl}(\Lambda)$, once some additional notation has been introduced.

Write

$$A = \coprod_1^s A_i \quad \text{(simple components)},$$

$$C = \coprod_1^s F_i, \quad F_i = \text{center of } A_i,$$

$$O = \coprod_1^s R_i, \quad R_i = \text{integral closure of } R \text{ in } F_i.$$

For each i, $1 \le i \le s$, let S_i be the set of all real primes P of F_i such that $(A_i)_P$ is not a full matrix algebra over the real field $(F_i)_P$ (we say that A_i *ramifies* at each such P). Let

$$F_i^+ = \{\alpha \in F_i : \alpha \ne 0, \alpha_P > 0 \text{ for all } P \in S_i\}.$$

Put

$$C^+ = \coprod_1^s F_i^+ .$$

Let us also introduce the idèle group

$$J(F_i) = \left\{ (a_P) \in \prod_P u((F_i)_P) : a_P \in u((R_i)_P) \text{ almost everywhere} \right\} ,$$

and let

$$J(C) = \coprod_1^s J(F_i) ,$$

the idèle group of C.

For each i, $1 \le i \le s$, there is a reduced norm map $\mathrm{nr} : A_i \to F_i$, which is roughly like a determinant. The image of $u(A_i)$ is precisely F_i^+ (this is the Hasse-Schilling-Maass Norm Theorem), while by a deep result of Wang, the kernel of nr (acting on $u(A_i)$) is precisely the commutator subgroup of $u(A_i)$. This reduced norm map extends to maps

$$\mathrm{nr} : (A_i)_P \to (F_i)_P , \quad 1 \le i \le s , \quad P = \text{maximal ideal of } R ,$$

and hence yields a homomorphism

$$\mathrm{nr} : J(A) \to J(C) .$$

It turns out that

$$\mathrm{nr}\, J(A) = J(C) , \quad \mathrm{nr}\, u(A) = C^+ , \quad \mathrm{nr}\, J'(A) = 1 ,$$

and the preceding theorem then gives

COROLLARY. *Let Λ be an R-order in a semisimple K-algebra A, where K is an algebraic number field. Then*

$$\mathrm{cl}(\Lambda) \cong J(C) \Big/ \left\{ C^+ \cdot \prod_P \mathrm{nr}\, u(\Lambda_P) \right\} .$$

In practice, this formula is rather hard to apply, since it is usually difficult to specify the group of units $u(\Lambda_P)$ and to determine its image $\mathrm{nr}\, u(\Lambda_P)$. One relatively easy case is that in which Λ is a maximal R-order. In this case, we may write $\Lambda = \coprod_1^s \Lambda_i$, where Λ_i is a maximal R_i-order in the central simple F_i-algebra A_i. Clearly

$$\mathrm{cl}(\Lambda) \cong \coprod \mathrm{cl}(\Lambda_i) ,$$

so it suffices to handle the case of maximal orders in central simple algebras.

Changing notation, let A be a simple algebra with center K, and let Λ be a maximal R-order in A. Then for each P, Λ_P is a maximal R_P-order in A_P, and it is known that

$$\operatorname{nr} u(\Lambda_P) = u(R_P) .$$

Thus we obtain

$$\operatorname{cl}(\Lambda) \cong J(K)\Big/\Big\{K^+ \cdot \prod_P u(R_P)\Big\} ,$$

where

$$K^+ = \{a \in K : a \neq 0,\ a_P > 0 \text{ at every infinite prime } P \text{ of } K \text{ at which } A \text{ ramifies}\} .$$

By means of the usual procedure for switching from idèles to ideals, we then obtain

$$\operatorname{cl}(\Lambda) \cong I(R)/P^+(R) ,$$

where

$I(R)$ = group of fractional R-ideals of K,

$P^+(R)$ = subgroup consisting of all principal ideals Ra, $a \in K^+$.

Thus, class groups of maximal orders are almost the same as ideal class groups of their centers.

3. Computations and K-theory

Suppose throughout that Λ is an R-order in a finite dimensional semisimple K-algebra A, where K is an algebraic number field. Let $\operatorname{cl}(\Lambda)$ denote the locally free class group of Λ. We begin by listing some relatively simple properties of such class groups:

(i) Let $\rho : \Lambda \to \Gamma$ be a homomorphism of R-orders in semisimple K-algebras. Then ρ induces a homomorphism

$$\rho_* : \operatorname{cl}(\Lambda) \to \operatorname{cl}(\Gamma) ,$$

given by

$$\rho_*[M] = [\Gamma \otimes_\Lambda M] , \quad M = \text{locally free rank } 1\ \Lambda\text{-lattice.}$$

This map is consistent with the maps obtained by using the idèle-theoretic formulas for class groups.

(ii) Given an R-order Λ in A, let Λ' be a maximal R-order in A containing Λ. Then the map $\operatorname{cl}(\Lambda) \to \operatorname{cl}(\Lambda')$ is surjective. We denote its kernel by $D(\Lambda)$, so there is an exact sequence of additive groups

$$0 \to D(\Lambda) \to \operatorname{cl}(\Lambda) \to \operatorname{cl}(\Lambda') \to 0 .$$

In practice, cl(Λ') can be calculated explicitly by the formulas given at the end of the preceding section. Indeed, Λ' splits into a direct sum of maximal orders in simple algebras, and thus cl(Λ') is a direct product of strict ideal class groups of rings of algebraic integers. Thus we may regard cl(Λ') as known, and concentrate on the determination of the "kernel group" $D(\Lambda)$. In practice, this is usually the difficult part of the problem. Even more difficult is the determination of the additive structure of cl(Λ) once $D(\Lambda)$ and cl(Λ') are known.

(iii) Let $\rho : \Lambda \to \Gamma$ be a homomorphism of R-orders as in (i), and let ρ_* be the induced "change of rings" map on class groups. Then

$$\rho_* : D(\Lambda) \to D(\Gamma) ,$$

so both cl( ) and D( ) are functorial.

This result escaped notice for some time; the simplest proof is based on

LEMMA. *Let M, N be locally free Λ-lattices, and let Λ' be a maximal order containing Λ. Then $\Lambda'M$ is stably isomorphic to $\Lambda'N$ if and only if there exists a finitely generated Λ-module X such that $M \oplus X \cong N \oplus X$.*

(iv) One of the more interesting "qualitative" results, proved by means of the idèle-theoretic formula for class groups, is as follows:

THEOREM. *Let $\Lambda = ZG$, where G is a finite p-group. Then $D(\Lambda)$ is also a p-group.*

Let us now indicate the connection between class groups and algebraic K-theory. For any ring Λ, let $P(\Lambda)$ be the category of finitely generated projective left Λ-modules. Let $K_0(\Lambda)$ be the abelian group generated by symbols $[M]$, $M \in P(\Lambda)$, one for each isomorphism class of modules, with relations

$$[M \oplus N] = [M] + [N] , \quad M, N \in P(\Lambda) .$$

One calls $K_0(\Lambda)$ the *projective class group* of Λ. For $M, N \in P(\Lambda)$, we have

$$[M] = [N] \text{ in } K_0(\Lambda) \Longleftrightarrow M \text{ is stably isomorphic to } N .$$

In particular, let $\Lambda = ZG$ be an integral group ring of a finite group G. The inclusion $ZG \subset QG$ induces a homomorphism, whose kernel is denoted by $\tilde{K}_0(ZG)$, the *reduced projective class group* of ZG. This group is of interest to algebraic topologists; for example, associated with a topological space with fundamental group G, there is a Swan-Wall invariant lying in $\tilde{K}_0(ZG)$ which measures whether the space has the same homotopy type as a finite complex.

A fundamental theorem due to Swan asserts:

Every projective ZG-lattice is locally free, that is, lies in the same genus as a free module $(ZG)^{(n)}$ for some n.

From this fact, it is a simple matter to set up an isomorphism

$$\mathrm{cl}(ZG) \cong \tilde{K}_0(ZG) .$$

The isomorphism is given by

$$[\Lambda\alpha] \to [\Lambda\alpha] - [\Lambda] , \quad \alpha \in J(A) ,$$

where $\Lambda = ZG$, $A = QG$. This explains to some extent the importance of the locally free class group introduced in §2.

The connection between class groups and K-theory has also provided an extremely powerful method for calculation of class groups. Let us start with a fibre product diagram of rings and ring homomorphisms:

$$\begin{array}{ccc} \Lambda & \longrightarrow & \Lambda_1 \\ \downarrow & & \downarrow f_1 \\ \Lambda_2 & \xrightarrow{f_2} & \bar{\Lambda} \end{array} .$$

This means that $\Lambda \cong \{(\lambda_1, \lambda_2) : \lambda_i \in \Lambda_i , f_1(\lambda_1) = f_2(\lambda_2)\}$. Milnor proved

THEOREM. *Suppose that either* f_1 *or* f_2 *is surjective. Then there is an exact (Mayer-Vietoris) sequence*

$$K_1(\Lambda) \to K_1(\Lambda_1) \oplus K_1(\Lambda_2) \to K_1(\bar{\Lambda}) \to K_0(\Lambda) \to K_0(\Lambda_1) \oplus K_0(\Lambda_2) \to K_0(\bar{\Lambda}) .$$

Further, if both f_1 *and* f_2 *are surjective, the exact sequence can be extended to the left thus:*

$$K_2(\Lambda) \to K_2(\Lambda_1) \oplus K_2(\Lambda_2) \to K_2(\bar{\Lambda}) \to K_1(\Lambda) \to \dots .$$

This suggests strongly that an analogous sequence should exist for class groups. Indeed, it is not hard to prove the following (due to Reiner and Ullom):

THEOREM. *Let* Λ *be an* R*-order in a semisimple* K*-algebra* A *, where* K *is an algebraic number field, and where no simple component of* A *is a totally definite quaternion algebra. Let*

$$\begin{array}{ccc} \Lambda & \longrightarrow & \Lambda_1 \\ \downarrow & & \downarrow f_1 \\ \Lambda_2 & \xrightarrow{f_2} & \bar{\Lambda} \end{array}$$

be a fibre product in which either f_1 *or* f_2 *is surjective, where* Λ_1 *and* Λ_2 *are*

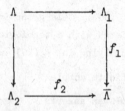

R-orders in semisimple K-algebras, and where $\bar{\Lambda}$ *is a finite ring. Let*

$$u^*(\Lambda_i) = f_i\{u(\Lambda_i)\}\ , \quad i = 1, 2\ .$$

Then there are exact sequences

$$\begin{array}{ccl} & & \nearrow^{\delta}\ \mathrm{cl}\,\Lambda \xrightarrow{\psi} \mathrm{cl}\,\Lambda_1 \oplus \mathrm{cl}\,\Lambda_2 \to 0 \\ 1 \to u^*(\Lambda_1)\cdot u^*(\Lambda_2) \to u(\bar{\Lambda}) & & \\ & & \searrow_{\delta}\ D(\Lambda) \xrightarrow{\psi} D(\Lambda_1) \oplus D(\Lambda_2) \to 0\ . \end{array}$$

The maps ψ are induced by the homomorphisms $\Lambda \to \Lambda_i$, $i = 1, 2$. The map δ is given thus: for each unit $u \in u(\bar{\Lambda})$, let

$$\Lambda u = \{(\lambda_1, \lambda_2) : \lambda_i \in \Lambda_i,\ f_1(\lambda_1)\cdot u = f_2(\lambda_2)\}\ ,$$

and set

$$\delta(u) = [\Lambda u]\ .$$

Finally,

$$u^*(\Lambda_1)\cdot u^*(\Lambda_2) = \{u_1u_2 : u_i \in u^*(\Lambda_i),\ i = 1, 2\}\ .$$

A slightly more complicated version of this theorem can be established for the case where A fails to satisfy the stated condition.

As a simple application of the above result, we evaluate $\mathrm{cl}(\Lambda)$, where $\Lambda = ZG$, G cyclic of prime order p . There is an identification $\Lambda \cong Z[x]/(x^p-1)$. Let

$$\Phi(x) = x^{p-1} + x^{p-2} + \ldots + x + 1\ , \quad R = Z[x]/(\Phi(x))\ ,$$

so we may identify R with the ring alg.int.$\{Q(\omega)\}$, where ω is a primitive pth root of 1 over Q . There is a fibre product diagram

$$\begin{array}{ccc} \Lambda & \longrightarrow & R \\ \downarrow & & \downarrow \\ Z & \longrightarrow & \bar{Z}\ , \end{array}$$

where $\bar{Z} = Z/pZ$, where $\Lambda \to R$ is the canonical map

$$Z[x]/(x^p-1) \to Z[x]/(\Phi(x))\ ,$$

and $\Lambda \to Z$ is given by

$$Z[x]/(x^p-1) \to Z[x]/(x-1)\ .$$

Thus we obtain an exact sequence

$$u(Z) \times u(R) \to u(\overline{Z}) \to \mathrm{cl}\ \Lambda \to \mathrm{cl}\ Z \oplus \mathrm{cl}\ R \to 0\ .$$

However, for each integer n prime to p, there is a unit $(\omega^n-1)/(\omega-1)$ of R with image $\overline{n}$ in $u(\overline{Z})$. This shows that $u(R)$ maps *onto* $u(\overline{Z})$. Since $\mathrm{cl}\ Z = 0$, this yields

$$\mathrm{cl}(ZG) \cong \mathrm{cl}(R) = \text{ideal class of group of } R\ ,$$

$$D(ZG) = 0\ .$$

One is tempted to try the case where G is cyclic of order p^2. Let $\Lambda = ZG$ and let ω_i be a primitive p^ith root of 1. Then there is an exact sequence

$$0 \to D(ZG) \to \mathrm{cl}(ZG) \to \mathrm{cl}\ Z[\omega_1] \oplus \mathrm{cl}\ Z[\omega_2] \to 0\ .$$

Kervaire and Murthy have shown:

If p is a regular odd prime (that is, if p does not divide $|\mathrm{cl}\ Z[\omega_1]|$), then $D(ZG)$ is an elementary abelian p-group on $(p-3)/2$ generators.

For further results on class groups, as well as for specific details of the theorems and propositions given in this article, we refer the reader to the texts and surveys listed below (see especially [5] and [6]).

References

[1] Charles W. Curtis, Irving Reiner, *Representation Theory of Finite Groups and Associative Algebras* (Pure and Applied Mathematics, **11**. Interscience [John Wiley & Sons], New York, London, 1962).

[2] T.Y. Lam and M.K. Siu, "K_0 and K_1 - an introduction to algebraic K-theory", *Amer. Math. Monthly* **82** (1975), 329-364.

[3] John Milnor, *Introduction to Algebraic K-Theory* (Annals of Mathematics Studies, **72**. Princeton University Press, Princeton, New Jersey; University of Tokyo Press, Tokyo; 1971).

[4] I. Reiner, *Maximal Orders* (London Mathematical Society Monographs, **5**. Academic Press [Harcourt Brace Jovanovich], London, New York, San Francisco, 1975).

[5] Irving Reiner, *Class Groups and Picard Groups of Group Rings and Orders* (Conference Board of the Mathematical Sciences Regional Conference Series, Mathematics, **26**. Amer. Math. Soc., Providence, Rhode Island, 1976).

[6] Irving Reiner, "Topics in integral representation theory", *Proc. Sao Paulo School of Algebra*, 1976 (Lecture Notes in Mathematics. Springer-Verlag, Berlin, Heidelberg, New York, to appear).

[7] Klaus W. Roggenkamp, *Lattices over Orders II* (Lecture Notes in Mathematics, 142. Springer-Verlag, Berlin, Heidelberg, New York, 1970).

[8] Klaus W. Roggenkamp and Verena Huber-Dyson, *Lattices over Orders I* (Lecture Notes in Mathematics, 115. Springer-Verlag, Berlin, Heidelberg, New York, 1970).

[9] Richard G. Swan, *K-theory of finite groups and orders* (notes by E. Graham Evans. Lecture Notes in Mathematics, 149. Springer-Verlag, Berlin, Heidelberg, New York, 1970).

[10] Stephen V. Ullom, "A survey of class groups of integral group rings", *Algebraic Number Fields: L-Functions and Galois Properties* (Proc. Sympos. Univ. Durham, Durham, 1975, 497-524. Academic Press, New York, London, 1977).

Department of Mathematics,
University of Illinois,
Urbana,
Illinois,
USA.

PROC. 18th SRI
CANBERRA 1978, 70-87.

20C10

INTEGRAL REPRESENTATIONS OF CYCLIC p-GROUPS

Irving Reiner

1. Introduction

Let ZG be the integral group ring of a finite group G. In order to classify all representations of G by matrices over Z, we must find a full set of non-isomorphic ZG-lattices. (By definition, a *ZG-lattice* is a left ZG-module with a finite free Z-basis.) It is easily seen that every ZG-lattice is expressible as a finite direct sum of indecomposable lattices. However, the Krull-Schmidt-Azumaya Theorem rarely holds for ZG-lattices. We are thus faced with three problems, listed in order of increasing difficulty:

(I) When is the number of isomorphism classes of indecomposable ZG-lattices finite?

(II) Find all indecomposable lattices.

(III) When are two direct sums of indecomposable lattices isomorphic?

The first problem was solved in 1962 by Jones [7], using results of Heller and Reiner [5, 6]; an independent solution was given by Berman and Gudivok [1]. The result is as follows:

THEOREM. *The number of isomorphism classes of indecomposable ZG-lattices is finite if and only if for each prime p dividing $|G|$, the p-Sylow subgroups of G are cyclic of order p or p^2.*

Problem (II) has been solved for G cyclic of order p or p^2, where p is prime (for the latter, see Reiner [12]). It has also been solved when G is dihedral of order $2p$, or more generally, dihedral of order pq (see [10] for references). There is also a solution due to Nazarova for the case where G is elementary abelian of type (2, 2), and also when G is the alternating group A_4, even though in these cases there are infinitely many indecomposable ZG-lattices.

The third problem is largely untouched, except for G cyclic of order p or p^2, or dihedral of order $2p$. We shall discuss in this article the case where G is cyclic of order p^2, though many of our results apply equally well to arbitrary cyclic p-groups.

2. Genus and extensions

For a prime p, let Z_p denote the p-adic completion of Z. Let M, N be ZG-lattices, where G is an arbitrary group. We say that M and N are in the same *genus* (notation: $M \vee N$) if $M_p \cong N_p$ for each prime p. It suffices to impose this condition for those primes p which divide $|G|$. In trying to find all ZG-lattices, the most fruitful approach is to begin by finding all genera of lattices, and then to classify all isomorphism classes within each genus. In some cases it may be extremely difficult to determine all genera, but then relatively easy to find all isomorphism classes within each genus. In other cases, and especially for G cyclic of order p^2, the difficulty lies in determining the isomorphism classes in each genus. In order to classify all ZG-lattices for arbitrary G, we first give a full set of genus invariants, and then seek extra invariants which will characterize the isomorphism classes within a given genus.

One of the main techniques for describing ZG-lattices is the use of extensions. Given a pair of lattices M, N, we consider all lattices X with

$$0 \to M \to X \to N \to 0$$

an exact sequence of ZG-lattices. Such X's are determined by the group $\mathrm{ext}^1_{ZG}(N, M)$, whose elements are called *extension classes*. Each extension class determines a single isomorphism class of X's. However, different elements of ext may possibly give isomorphic lattices X. In some cases, we can say precisely when this occurs:

LEMMA. *Let Λ be any ring, and let M, N be Λ-modules such that* $\hom_\Lambda(M, N) = 0$. *Let $\xi_1, \xi_2 \in \mathrm{ext}^1_\Lambda(N, M)$, and let X_i be a Λ-module determined by ξ_i, $i = 1, 2$. Then $X_1 \cong X_2$ if and only if*

$$\gamma\xi_1 = \xi_2\delta \quad \textit{for some} \quad \gamma \in \mathrm{aut}(M)\,, \quad \delta \in \mathrm{aut}(N)\,.$$

Proof. Given

$$\begin{array}{ccccccccc} \xi_1 : & 0 \to M & \xrightarrow{f} & X_1 & \longrightarrow & N \to 0 \\ & & & \Big\downarrow \varphi & & \\ \xi_2 : & 0 \to M & \longrightarrow & X_2 & \xrightarrow{g} & N \to 0 , \end{array}$$

where φ is a Λ-isomorphism, we obtain a Λ-homomorphism $g\varphi f : M \to N$. Thus $g\varphi f = 0$ by hypothesis, whence φ induces maps $\gamma \in \text{end}_\Lambda(M)$, $\delta \in \text{end}_\Lambda(N)$. The resulting commutative diagram shows that $\gamma\xi_1 = \xi_2\delta$. Since φ is an isomorphism, so are γ and δ. The argument can be reversed, so the lemma is proved.

COROLLARY. *Let M, N be Λ-modules such that* $\text{hom}_\Lambda(M, N) = 0$, *where Λ is an arbitrary ring. Then there is a bijection between the set of isomorphism classes of extensions of N by M, and the set of orbits of* $\text{ext}^1_\Lambda(N, M)$ *under the actions of* $\text{aut}_\Lambda(N)$ *from the right and* $\text{aut}_\Lambda(M)$ *from the left.*

Let us now restrict our attention to the case where Λ is an R-order in a finite-dimensional semisimple K-algebra A, where R is a Dedekind ring whose quotient field K is an algebraic number field. We may choose a non-empty set $S(\Lambda)$ of maximal ideals P of R, such that for each $P \notin S(\Lambda)$, the P-adic completion Λ_P is a maximal R_P-order in A_P. Let M, N be Λ-lattices; then M, N are in the same genus if and only if $M_P \cong N_P$ for all $P \in S(\Lambda)$. (If $\Lambda = RG$, where G is a finite group, then we may choose $S(\Lambda)$ to be any non-empty finite set which includes all P's which divide $|G|$.)

Now consider $\text{ext}^1_\Lambda(N, M)$, where M, N are arbitrary Λ-lattices; we abbreviate this as $\text{ext}(N, M)$. Then (see [2, (85.22)]) $\text{ext}(N, M)$ is a finitely generated torsion R-module, and

$$\text{ext}^1_\Lambda(N, M) \cong \coprod_{P \in S(\Lambda)} \text{ext}^1_{\Lambda_P}(N_P, M_P) .$$

In particular, this shows that (up to isomorphism) $\text{ext}(N, M)$ depends only upon the genera of N and M, rather than on the isomorphism classes of N and M.

We need an analogue of Schanuel's Lemma, due to Roiter:

LEMMA. *Let X, X', Y, Y' be Λ-lattices, and let T be an R-torsion Λ-module such that $T_P = 0$ for each $P \in S(\Lambda)$. Suppose that there exists a pair of Λ-exact sequences*

$$0 \to X' \to X \xrightarrow{f} T \to 0 , \quad 0 \to Y' \to Y \xrightarrow{g} T \to 0 .$$

Then there is a Λ*-isomorphism*

$$X \oplus Y' \cong X' \oplus Y .$$

Proof. Let W be the pullback of the pair of maps f, g. Then there is a commutative diagram of Λ-modules, with exact rows and columns:

$$\begin{array}{ccccccccc} & & & & 0 & & 0 & & \\ & & & & \downarrow & & \downarrow & & \\ & & & & X' & \xrightarrow{1} & X' & & \\ & & & & \downarrow & & \downarrow & & \\ 0 & \to & Y' & \longrightarrow & W & \longrightarrow & X & \to & 0 \\ & & {\scriptstyle 1}\downarrow & & \downarrow & & \downarrow{\scriptstyle f} & & \\ 0 & \to & Y' & \longrightarrow & Y & \xrightarrow{g} & T & \to & 0 \\ & & & & \downarrow & & \downarrow & & \\ & & & & 0 & & 0 & & \end{array}$$

The process of forming P-adic completions preserves commutativity and exactness, since R_P is flat as R-module. At each $P \in S(\Lambda)$ we have $T_P = 0$, so $W_P \cong X_P \oplus Y_P$. Therefore both of the Λ-exact sequences

$$0 \to X' \to W \to Y \to 0 , \quad 0 \to Y' \to W \to X \to 0$$

are split at each $P \in S(\Lambda)$. On the other hand, for $P \notin S(\Lambda)$ we know that Λ_P is a maximal order, so both of the Λ_P-lattices Y_P and X_P are Λ_P-projective. Thus both of the above sequences split at each $P \notin S(\Lambda)$. This shows that the sequences split at *every* maximal ideal P of R, and so they split globally:

$$W \cong X' \oplus Y , \quad W \cong X \oplus Y' ,$$

which proves the lemma.

We may use this lemma to derive various identities concerning direct sums of lattices. Let us introduce some convenient notation for describing such formulas. Let M, N be Λ-lattices, and let $\xi \in \text{ext}(N, M)$ determine a Λ-lattice X (unique up to isomorphism). Each $\gamma \in \text{end}_\Lambda(M)$ acts on $\text{ext}(N, M)$, so we can form the extension class $\gamma\xi$, and denote by ${}_\gamma X$ the lattice determined by $\gamma\xi$. The relation between X and ${}_\gamma X$ is described by the commutative diagram with Λ-exact rows:

$$\begin{array}{lccccccccc} \xi : & 0 & \to & M & \xrightarrow{f} & X & \longrightarrow & N & \to & 0 \\ & & & {\scriptstyle \gamma}\downarrow & & {\scriptstyle \gamma'}\downarrow & & \downarrow{\scriptstyle 1} & & \\ \gamma\xi : & 0 & \to & M & \longrightarrow & {}_\gamma X & \longrightarrow & N & \to & 0 , \end{array}$$

and indeed ${}_\gamma X$ is the pushout of the pair of Λ-homomorphisms f, γ. In particular, suppose that $\gamma \in \text{end}_\Lambda(M)$ is such that

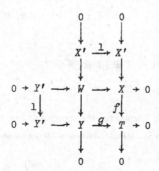

$$\gamma \in \operatorname{aut}_{\Lambda_P}(M_P) \quad \text{for each} \quad P \in S(\Lambda) .$$

Then $X_P \cong ({}_\gamma X)_P$ for $P \in S(\Lambda)$, so ${}_\gamma X \vee X$.

EXCHANGE FORMULA. *Let* M, N *be* Λ*-lattices, and let* $\gamma \in \operatorname{end}_\Lambda(M)$ *be such that* $\gamma \in \operatorname{aut}(M_P)$ *for each* $P \in S(\Lambda)$. *Let* $\xi_1, \xi_2 \in \operatorname{ext}(N, M)$ *determine* Λ*-lattices* X, Y , *respectively. Then*

$$X \oplus {}_\gamma Y \cong {}_\gamma X \oplus Y$$

as Λ*-modules.*

Proof. Applying the Snake Lemma to the commutative diagram above, we have $\operatorname{cok} \gamma \cong \operatorname{cok} \gamma'$, $\ker \gamma \cong \ker \gamma'$. But $\ker \gamma = 0$ since $\ker \gamma_P = 0$ for $P \in S(\Lambda)$. Thus we obtain a Λ-exact sequence

$$0 \to X \to {}_\gamma X \to \operatorname{cok} \gamma \to 0 ,$$

with

$$(\operatorname{cok} \gamma)_P = 0 \quad \text{for all} \quad P \in S(\Lambda) .$$

A corresponding sequence holds for $Y, {}_\gamma Y$. The desired result then follows from the preceding lemma.

In the same manner, we obtain

ABSORPTION FORMULA. *Keeping the above notation and hypotheses, there is a* Λ*-isomorphism*

$${}_\gamma X \oplus M \cong X \oplus M .$$

Proof. We use the pair of exact sequences

$$0 \to X \to {}_\gamma X \to \operatorname{cok} \gamma \to 0 , \quad 0 \to M \xrightarrow{\gamma} M \to \operatorname{cok} \gamma \to 0 .$$

We have seen earlier that the group $\operatorname{ext}^1_\Lambda(N, M)$ depends only upon the genera of the Λ-lattices M and N . In particular, if $M' \vee M$ and $N' \vee N$, where M', N' are Λ-lattices, then

$$\operatorname{ext}^1_\Lambda(N', M') \cong \operatorname{ext}^1_\Lambda(N, M) .$$

It seems likely that there are as many isomorphism classes of extensions of N by M , as extensions of N' by M' . We may prove this under some mild restrictive hypothesis:

PROPOSITION. *Let* Λ *be an* R*-order in a semisimple* K*-algebra* A , *where* K *is an algebraic number field. Let* M, M', N, N' *be* Λ*-lattices such that* $M' \vee M$,

$N' \vee N$, and suppose that the Λ-lattice $M \oplus N$ satisfies the Eichler condition (that is, no simple component of $\mathrm{end}_A(KM \oplus KN)$ is a positive definite quaternion algebra Then there is a bijection between the set of isomorphism classes of extensions X of N by M, and the corresponding set of extensions X' of N' by M'.

Proof (Outline). We sketch the proof for the case in which we change only one of the "variables", say M. Thus, let $M' \vee M$, $N' = N$. Since $M' \vee M$, we can find a Λ-monomorphism $\varphi : M \to M'$ with $\operatorname{cok} \varphi$ an R-torsion Λ-module for which $(\operatorname{cok} \varphi)_P = 0$ for each $P \in S(\Lambda)$, where as above, $S(\Lambda)$ is any finite non-empty set of maximal ideals P of R such that

$$S(\Lambda) \supseteq \{P : \Lambda_P \neq \text{maximal } R_P\text{-order in } A_P\} .$$

Now let X be an extension of N by M, and consider the diagram

$$\begin{array}{ccccccc} 0 \to & M & \xrightarrow{f} & X & \to & N & \to 0 \\ & \varphi\downarrow & & & & & \\ & M' & & & & & \end{array}$$

Define X' to be the pushout of the pair of maps (f, φ), so we obtain a commutative diagram

$$\begin{array}{ccccccc} 0 \to & M & \xrightarrow{f} & X & \longrightarrow & N & \to 0 \\ & \varphi\downarrow & & \downarrow & & 1\downarrow & \\ 0 \to & M' & \longrightarrow & X' & \longrightarrow & N & \to 0 . \end{array}$$

and

$$0 \to X \to X' \to \operatorname{cok} \varphi \to 0$$

is exact. This implies that $X' \vee X$.

Thus, once φ is fixed, each X gives rise to an X'. If also Y is an extension of N by M, with corresponding lattice Y' defined in an analogous manner, then $Y' \vee Y$, and

$$0 \to Y \to Y' \to \operatorname{cok} \varphi \to 0$$

is exact. By Roiter's "Schanuel Lemma", this implies that

$$X \oplus Y' \cong Y \oplus X' .$$

Suppose now that $X \cong Y$; then X, Y, X', Y' are in the same genus, and clearly $KX \cong KM \oplus KN$. The Jacobinski Cancellation Theorem asserts that from the isomorphism

$$X \oplus Y' \cong X \oplus X'$$

we may conclude that $Y' \cong X'$, provided we assume that KX satisfies the Eichler condition. Since this is part of our hypothesis, we obtain

$$X \cong Y \text{ if and only if } X' \cong Y' .$$

To conclude the proof, we need only show that every X' comes from some X; this is easily done by finding an embedding $\psi : M' \to M$ such that $\psi\varphi$ is congruent to 1 modulo a high power of P, for each $P \in S(\Lambda)$. See [13] for details.

REMARK. It seems likely that one can omit the hypothesis that $KM \oplus KN$ satisfy the Eichler condition.

3. Cyclic p-groups

Let p be prime, and let G be a cyclic group of order p^2. We shall show how to classify ZG-lattices. Many of the calculations below apply equally well to the case where G is an arbitrary cyclic p-group, but we cannot hope for a complete classification in the general case, since the representation problem involved is of wild type for $p \geq 5$.

For each i, we put

$$\Lambda_i = Z[x]/\left(x^{p^i}-1\right), \quad R_i = Z[\omega_i], \quad K_i = \text{quotient field of } R_i,$$

where ω_i is a primitive p^i-th root of 1 over Q. If $\Phi_i(x)$ is the cyclotomic polynomial of order p^i and degree $\varphi(p^i)$, there is an isomorphism

$$R_i \cong Z[x]/\left(\Phi_i(x)\right),$$

which we treat as an identification. Let G be cyclic of order p^2, and identify ZG with the ring Λ_2. For $i = 0, 1, 2$, the Dedekind ring R_i is a quotient ring of Λ_2, so each R_i-lattice may also be viewed as a ZG-lattice.

Given any ZG-lattice M, define

$$L = \{m \in M : (x^p-1)m = 0\}.$$

Then L is a Λ_1-lattice in M, and it is easily checked that M/L is an R_2-lattice. Thus we obtain a ZG-exact sequence

$$0 \to L \to M \to N \to 0$$

in which L is a Λ_1-lattice, N an R_2-lattice, and we need to classify extensions of N by L. Since $(x^p-1)L = 0$ it follows that $\Phi_2(x)$ acts as p on L. But $\Phi_2(x)\cdot N = 0$, so $\hom_{ZG}(L, N) = 0$. This shows that M determines L, N up to isomorphism, and in order to classify all M's, we need only solve the following problems:

(i) Classify all R_2-lattices N;

(ii) Classify all Λ_1-lattices L ;

(iii) Compute the orbits of $\mathrm{ext}^1_{ZG}(N, L)$ under the actions of aut(N) and aut(L) .

Now R_2 is a Dedekind ring, since $R_2 = \mathrm{alg.int.}\{K_2\}$. Thus we may apply Steinitz's Theorem to find all R_2-lattices. If N is an R_2-lattice of rank d (where $d = (K_2N : K_2)$), then

$$N \cong \coprod_1^d \underline{\underline{b}}_i \ , \quad \underline{\underline{b}}_i = \text{nonzero ideal in } R_2 \ .$$

The lattice N is determined up to isomorphism by its rank d and its *Steinitz class*, defined as the ideal class of the product $\prod \underline{\underline{b}}_i$. We note also that $N \vee R_2^{(d)}$, so the rank d determines the *genus* of the R_2-lattice N .

The classification of Λ_1-lattices is slightly more complicated, but in fact can be obtained as a special case of the calculations given below. Since this classification has been known for about 20 years by now, we merely quote the result:

For each nonzero ideal $\underline{\underline{a}}$ of R_1 , let $E(\underline{\underline{a}})$ denote the Λ_1-lattice which is a non-split extension of $\underline{\underline{a}}$ by the trivial Λ_1-lattice Z :

$$0 \to Z \to E(\underline{\underline{a}}) \to \underline{\underline{a}} \to 0 \ .$$

It is easily checked that for each $\underline{\underline{a}}$, there is only one such $E(\underline{\underline{a}})$, up to isomorphism. Then every Λ_1-lattice L is of the form

$$L \cong Z^{(a)} \oplus \coprod_1^b \underline{\underline{a}}_i \oplus \coprod_{b+1}^{b+c} E(\underline{\underline{a}}_i) \ .$$

The isomorphism class of L is completely determined by its genus invariants a, b, c , and by its Steinitz class, defined as the R_1-ideal class of the product

$$\prod_1^{b+c} \underline{\underline{a}}_i \ .$$

Thus, we know L and N , and it remains to calculate the orbits of ext(N, L) under the actions of aut(N) and aut(L) . We are dealing with a situation in which all of the modules involved satisfy the Eichler condition, and so in calculating these orbits, we may replace N by any lattice in the genus of N , and likewise we may replace L by any lattice in its genus. Hence we may choose

$$N = R_2^{(d)} \ , \quad L = Z^{(a)} \oplus R_1^{(b)} \oplus \Lambda_1^{(c)} \ ,$$

since, for each $\underline{\underline{a}}$, $E(\underline{\underline{a}})$ lies in the same genus as Λ_1 .

Let us calculate $\text{ext}^1_{ZG}(L, N)$. We have

$$\text{ext}(L, N) \cong \text{ext}\left(L, R_2^{(d)}\right) \cong \{\text{ext}(L, R_2)\}^{(d)} .$$

There is a Λ_2-exact sequence

$$0 \to \Phi_2(x)\Lambda_2 \xrightarrow{i} \Lambda_2 \to R_2 \to 0 ,$$

giving rise to an exact sequence

$$\text{hom}(\Lambda_2, L) \xrightarrow{i^*} \text{hom}(\Phi_2(x)\Lambda_2, L) \to \text{ext}^1_{\Lambda_2}(R_2, L) \to 0 .$$

But

$$\text{hom}(\Phi_2(x)\Lambda_2, L) \cong L ,$$

by means of the isomorphism which carries an element $f \in \text{hom}(\Phi_2(x)\Lambda_2, L)$ onto the element $f(\Phi_2(x)) \in L$, which may be arbitrary since $(x^p-1)L = 0$. In this isomorphism, the image of i^* is precisely $\Phi_2(x)\cdot L$, which coincides with pL . Thus

$$\text{ext}(R_2, L) \cong L/pL = \bar{L} \text{ (say)},$$

and so we have

$$\text{ext}(N, L) \cong \bar{L}^{(d)} ,$$

where L is as given above. Note that

$$\bar{L} \cong \bar{Z}^{(a)} \oplus \bar{R}_1^{(b)} \oplus \bar{\Lambda}_1^{(c)}$$

where bars denote reduction mod p . Thus $\text{ext}(N, L)$ is known explicitly, and we are now ready to consider the difficult question of determining the orbits in $\text{ext}(N, L)$.

4. Orbits and cyclotomic units

We consider next the special case in which

$$N = R_2^{(d)} , \quad L = R_1^{(b)}, \quad \text{ext}(N, L) \cong \bar{R}_1^{b\times d} ,$$

where $\bar{R}_1 = R_1/pR_1$ and where $\bar{R}_1^{b\times d}$ denotes the set of all $b \times d$ matrices over $\bar{R}_1$. Clearly

$$\text{aut}(N) \cong \text{GL}(d, R_2) , \quad \text{aut}(L) \cong \text{GL}(b, R_1) .$$

Furthermore, there are ring surjections $R_1 \to \overline{R}_1$, $R_2 \to \overline{R}_1$, by means of which these automorphism groups act on $\overline{R}_1^{b\times d}$. Let us call two matrices $\xi, \xi' \in \overline{R}_1^{b\times d}$ *strongly equivalent* (notation: $\xi' \approx \xi$) if

$$\xi' = \alpha\xi\beta \quad \text{for some} \quad \alpha \in \mathrm{GL}(b, R_1) \;,\; \beta \in \mathrm{GL}(d, R_2) \;.$$

Then the orbits of ext(N, L) under the actions of aut(N) and aut(L) are precisely the strong equivalence classes of $\overline{R}_1^{b\times d}$.

(Before discussing these strong equivalence classes, we may consider briefly another problem of the same nature. Let $e > 0$, and consider matrices ξ, ξ' over the ring Z/p^eZ ; write $\xi' \approx \xi$ if $\xi' = \alpha\xi\beta$ for some $\alpha, \beta \in \mathrm{GL}(Z)$. It is a natural question to ask for the strong equivalence classes of such matrices, but this question does not seem to have been considered before.)

Returning to our case of matrices over $\overline{R}_1^{b\times d}$, we note that

$$\overline{R}_1 = R_1/pR_1 = Z[x]/\big(p, \Phi_1(x)\big) \cong \overline{Z}[x]/\big(\Phi_1(x)\big)$$
$$\cong \overline{Z}[x]/\big(\lambda^{p-1}\big) \;, \text{ where } \lambda = 1 - x \;.$$

Thus $\overline{R}_1$ is a local principal ideal ring, whose nonzero ideals are precisely $\left\{\lambda^i\overline{R}_1 : 0 \le i \le p-2\right\}$. We may use the usual theory of elementary divisors for matrices over $\overline{R}_1$, and we have at once

LEMMA. *If* $\xi, \xi' \in \overline{R}_1^{b\times d}$ *are strongly equivalent, then*

$$\text{el.div.}(\xi) = \text{el.div.}(\xi') \;.$$

Thus, these elementary divisors are invariants of the orbit of ξ . Furthermore, starting with any $\xi \in \overline{R}_1^{b\times d}$, we may apply elementary row and column operations to ξ ; by definition, an elementary row operation is one which interchanges two rows of ξ , or adds to one row an $\overline{R}_1$-multiple of another row. Since $\overline{R}_1$ is a local ring, we can diagonalize ξ by means of such elementary row and column operations. However, these operations are images of corresponding operations over R_1 or R_2 , in view of the ring surjections $R_1 \to \overline{R}_1$, $R_2 \to \overline{R}_1$. Thus (if $b \le d$) we have

$$\xi \approx [D \; 0] \;,\; D = \mathrm{diag}\left(\lambda^{k_1}u_1, \ldots, \lambda^{k_b}u_b\right) ,$$

where

$$0 \le k_1 \le \ldots \le k_b \le p-1 \;,\; u_i \in u\big(\overline{R}_1\big) \;,$$

and $u(\overline{R}_1)$ denotes the group of units of $\overline{R}_1$. Clearly

$$\text{el.div.}(\xi) = \{\lambda^{k_1}, \ldots, \lambda^{k_b}\},$$

so the set of exponents $\{k_i\}$ is uniquely determined by the class of ξ .

On the other hand, for each $u \in u(\overline{R}_1)$ we may express $\begin{bmatrix} u & 0 \\ 0 & u^{-1} \end{bmatrix}$ as a product of elementary matrices. This implies that

$$\xi \approx [D\ 0] , \quad D = \text{diag}\left(\lambda^{k_1}, \ldots, \lambda^{k_{b-1}}, \lambda^{k_b}u\right)$$

where

$$0 \le k_1 \le \ldots \le k_b \le p\text{-}1 , \text{ and } u \in u(\overline{R}_1) .$$

If $b \ne d$, the same procedure permits us to eliminate u altogether, and we obtain

LEMMA. *Let* $\xi, \xi' \in \overline{R}_1^{b\times d}$ *, where* $b \ne d$ *. Then*

$$\xi' \approx \xi \textit{ if and only if } \text{el.div.}(\xi) = \text{el.div.}(\xi') .$$

Turning next to the case where $\xi \in \overline{R}_1^{b\times b}$, we may first diagonalize ξ , and are then faced with the question of deciding when two diagonal matrices are strongly equivalent. Let

$$\xi \approx D = \text{diag}\left(\lambda^{k_1}, \ldots, \lambda^{k_{b-1}}, \lambda^{k_b}u\right) , \quad u \in u(\overline{R}_1) ,$$

where

$$0 \le k_1 \le \ldots \le k_b \le p\text{-}1 .$$

Once the $\{k_i\}$ are given, it is clear that the strong equivalence class of ξ is unchanged if u is replaced by c_1uc_2 , where $c_i \in u(R_i)$, $i = 1, 2$. Furthermore, u may be changed modulo λ^{p-1-k_b} without affecting the matrix D .

Let us define for $0 \le t \le p\text{-}1$,

$$U_t = \left\{\text{units of } \overline{R}_1/\lambda^t\overline{R}_1\right\} \Big/ \left\{\text{image of } u(R_1) \cdot \text{image of } u(R_2)\right\} .$$

Note that

$$\overline{R}_1/\lambda^t\overline{R}_1 \cong \overline{Z}[\lambda]/(\lambda^t) ,$$

and that both $u(R_1)$ and $u(R_2)$ map into $u\left(\overline{R}_1/\lambda^t\overline{R}_1\right)$ by composition of maps

$$R_i \to \overline{R}_1 \to \overline{R}_1/\lambda^t\overline{R}_1 \;, \quad i = 1, 2 \;, \quad 0 \le t \le p-1 \;.$$

PROPOSITION. *Let* $\xi \in \overline{R}_1^{b\times b}$ *, and let*

$$\xi \approx \mathrm{diag}\left(\lambda^{k_1}u_1, \ldots, \lambda^{k_b}u_b\right) \;, \quad u_i \in u(\overline{R}_1) \;,$$

where

$$\mathrm{el.div.}(\xi) = \{\lambda^{k_1}, \ldots, \lambda^{k_b}\} \;, \quad 0 \le k_i \le p-1 \;.$$

Let

$$u(\xi) = \textit{image of } \prod_1^b u_i \textit{ in } U_t \;,$$

where

$$t = p - 1 - \max\{k_i\} \;.$$

Then el.div.(ξ) *and* $u(\xi)$ *are invariants of the strong equivalence class of* ξ *, and completely determine this class.*

The only part requiring proof is the fact that $u(\xi)$ is indeed an invariant of the strong equivalence class of ξ . This fact is an easy consequence of a rather interesting fact about determinants:

LEMMA. *Let* Γ *be any commutative ring, and let*

$$D = \mathrm{diag}\left(\gamma_1, \ldots, \gamma_m\right) \;, \quad \gamma_i \in \Gamma \;.$$

Suppose that γ_m *is a multiple of each* γ_i *, say*

$$\gamma_m = r_1\gamma_1 = r_2\gamma_2 = \ldots = r_{m-1}\gamma_{m-1} \;, \quad r_i \in \Gamma \;.$$

Let $X, Y \in \Gamma^{m\times m}$ *be matrices such that*

$$XD = DY \;.$$

Then

$$(\det X)\gamma_m = (\det Y)\gamma_m$$

holds true in Γ .

Let us translate the above results into statements about lattices. Let $\underline{\underline{a}}$ be a nonzero ideal of R_1 , $\underline{\underline{b}}$ a nonzero ideal of R_2 ; we have seen that

$$\mathrm{ext}^1_{ZG}(\underline{\underline{b}}, \underline{\underline{a}}) \cong \underline{\underline{a}}/p\underline{\underline{a}} \cong \overline{R}_1 \;,$$

and we fix such an isomorphism once and for all. For $\lambda^k u \in \overline{R}_1$, where $0 \le k \le p-2$

and $u \in u(\overline{R}_1)$, let us denote by $(\underline{a}, \underline{b}; \lambda^k u)$ the ZG-lattice which is an extension of $\underline{b}$ by $\underline{a}$ with extension class whose image in $\overline{R}_1$ is $\lambda^k u$. Then we obtain

THEOREM. *Let M be any ZG-lattice which is an extension of an R_2-lattice N by an R_1-lattice L . Then M is a direct sum of indecomposable ZG-lattices of the following types:*

(i) ideals $\underline{a}$ of R_1,

(ii) ideals $\underline{b}$ of R_2,

(iii) nonsplit extensions $(\underline{a}, \underline{b}; \lambda^k u)$ of $\underline{b}$ by $\underline{a}$.

Furthermore, a full set of isomorphism invariants of a direct sum

$$\coprod_1^r \left(\underline{a}_i, \underline{b}_i; \lambda^{k_i} u_i\right) \oplus \coprod_{r+1}^{r+s} \underline{a}_j \oplus \coprod_{r+1}^{r+t} \underline{b}_j$$

is as follows:

(i) the genus invariants: r, s, t , and the set of exponents $\{\lambda^{k_i}\}$,

(ii) the R_1-ideal class of $\prod_1^{r+s} \underline{a}_i$ and the R_2-ideal class of $\prod_1^{r+t} \underline{b}_j$, and

(iii) in the special case where $s = 0$ and $t = 0$, the image of $u_1 \ldots u_r$ in U_t , where

$$t = p - 1 - \max\{k_i\} .$$

We remark that the same type of analysis can be carried out to classify all extensions of R_i-lattices by locally free E-lattices, where E is any factor ring of Λ_{i-1} , and where i is any positive integer. In particular, we can find all extensions of R_2-lattices by locally free Λ_1-lattices, a step which is vital for the eventual classification of all ZG-lattices when G is cyclic of order p^2 .

To conclude this section, let us consider the finite abelian group U_t defined above where $0 \le t \le p-1$. We note first that there is a commutative diagram

$$\begin{array}{ccc} u(R_2) & \xrightarrow{\text{norm}} & u(R_1) \\ & \searrow & \downarrow \\ & & u(\overline{R}_1) \end{array} ,$$

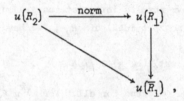

which implies at once that the image of $u(R_2)$ in $u\left(\overline{R}_1/\lambda^t\overline{R}_1\right)$ lies in the image of $u(R_1)$ in $u\left(\overline{R}_1/\lambda^t\overline{R}_1\right)$. Thus we obtain

$$U_t = \{\text{units of } \overline{Z}[\lambda]/(\lambda^t)\}/\text{image of } u(R_1)\ , \quad 0 \le t \le p-1 .$$

Now the units of the local ring $\overline{Z}[\lambda]/(\lambda^t)$ are easily determined; they are of the form $\sum_0^{t-1} a_i\lambda^i$, $a_i \in \overline{Z}$, $a_0 \neq 0$. The image of $u(R_1)$ is much harder to calculate, and by the work of Kervaire-Murthy [9] and Galovich [4], we obtain

THEOREM. *(i) Let* p *be a regular prime, that is, suppose* p *does not divide* $|\mathrm{cl}\ R_1|$. *Then for* $0 \le t \le p-1$, U_t *is an elementary abelian* p*-group on* $f(t)$ *generators, where* $f(t)$ *is the number of odd integers among* $3, 5, \ldots, t-1$. *(Thus* $f(t) = 0$ *if* $t \le 3$.*)*

(ii) Let p *be a properly irregular prime, that is,*

$$|\mathrm{cl}\ R_1| \equiv 0 \bmod p\ , \quad \left|\mathrm{cl}\ Z\left[\omega_1+\omega_1^{-1}\right]\right| \not\equiv 0 \pmod p) .$$

Let $\delta(k)$ *be the number of Bernoulli numbers among* $B_1, B_2, \ldots, B_k$ *whose numerators are divisible by* p. *Then* U_t *is an elementary abelian* p*-group on* $g(t)$ *generators, where*

$$g(t) = \left[\frac{t-2}{2}\right] + \delta\left[\frac{t-1}{2}\right], \quad 0 \le t \le p-1 ,$$

and where we interpret the greatest integer function $[(t-2)/2]$ *as* 0 *whenever* $t < 2$.

As far as is now known, every prime p is either regular or properly irregular.

In the classification of extensions of R_2-lattices by locally free Λ_1-lattices, one must also deal with the abelian p-group

$$U_p = u\left(\overline{Z}[\lambda]/(\lambda^p)\right)/\text{images of } u(R_2) \text{ and } u(\Lambda_1) .$$

This can be analysed in the same manner, and a theorem analogous to the above gives the structure of U_p.

5. Final classification

Let us now consider an arbitrary ZG-lattice M; we have seen that there is an exact sequence

$$0 \to L \to M \to N \to 0$$

in which L is a Λ_1-lattice and N an R_2-lattice. A full set of isomorphism invariants of M is the isomorphism class of L, the isomorphism class of N, and an orbit in $\operatorname{ext}(N, L)$. However, it is cumbersome to describe the invariants which specify such an orbit, and we adopt a somewhat different approach.

Let $\hat{Z}$ denote the p-adic completion of Z, $\hat{M}$ that of M. Two ZG-lattices M, N are in the same genus if $\hat{M} \cong \hat{N}$ as $\hat{Z}G$-modules. Furthermore, M is decomposable if and only if $\hat{M}$ is decomposable; for suppose that $\hat{M} = X \oplus Y$, where X, Y are $\hat{Z}G$-modules. Since every simple QG-module remains simple under extension of Q to its p-adic completion $\hat{Q}$, it follows that $\hat{Q}X = \hat{Q}V$ for some QG-module V. Put $M_0 = V \cap X$, a ZG-lattice in V for which $\hat{M}_0 = X$. In the same manner, we find a ZG-lattice M_1 with $\hat{M}_1 = Y$, so now

$$M \vee (M_1 \oplus M_2) .$$

But then M is also decomposable, since decomposability of a ZG-lattice depends only upon its genus. (This argument is due to Heller, in a more general format.)

In order to determine those orbits in $\operatorname{ext}(N, L)$ which correspond to indecomposable lattices, it thus suffices to consider the question locally rather than globally, and to work with $\hat{Z}G$-lattices rather than ZG-lattices. This local argument may be found in detail in [5, 6], and we merely record the consequences:

THEOREM. *Every indecomposable ZG-lattice is in the genus of exactly one of the following $4p + 1$ indecomposable lattices:*

$$\begin{cases} Z,\ R_1,\ R_2,\ \Lambda_1,\ (Z, R_2; 1), & \\ (\Lambda_1, R_2; \lambda^k), & 0 \le k \le p-1, \\ (Z \oplus \Lambda_1, R_2; 1 \oplus \lambda^k), & 1 \le k \le p-2, \\ (R_1, R_2; \lambda^k), & 0 \le k \le p-2, \\ (Z \oplus R_1, R_2; 1 \oplus \lambda^k), & 0 \le k \le p-2. \end{cases}$$

Here, $(Z, R_2; 1)$ denotes an extension of R_2 by Z corresponding to the extension class $1 \in \overline{Z}$, using the isomorphism

$$\operatorname{ext}(R_2, Z) \cong \overline{Z} = Z/pZ .$$

Similar definitions hold for the other lattices in the list.

(Note that for $p > 2$, there are indecomposable ZG-lattices of Z-rank $p^2 + 1$, namely those of the form $(Z \oplus \Lambda_1, R_2; 1 \oplus \lambda^k)$, where

$$1 \oplus \lambda^k \in \overline{Z} \oplus \overline{\Lambda}_1 \cong \text{ext}(R_2, Z \oplus \Lambda_1) ,$$

with bars denoting reduction mod p .)

From the above list we may obtain *all* indecomposable ZG-lattices by means of the following changes:

(i) Replace R_1 by $\underline{\underline{a}}$, with $\underline{\underline{a}}$ ranging over a full set of representatives of the ideal classes of R_1 .

(ii) Replace Λ_1 by $E(\underline{\underline{a}})$, with $\underline{\underline{a}}$ as above, where $E(\underline{\underline{a}})$ is a nonsplit extension of $\underline{\underline{a}}$ by Z .

(iii) Replace R_2 by $\underline{\underline{b}}$, with $\underline{\underline{b}}$ ranging over a full set of representatives of the ideal classes of R_2 .

(iv) Replace λ^k by $\lambda^k u$, with u ranging over some suitably chosen subset of the group of units of $\overline{R}_1$ or $\overline{\Lambda}_1$, where bars denote reduction mod p .

The specific description of the range of u in (iv) is somewhat complicated, though quite explicit, and involves the abelian p-groups U_t defined in the preceding section (see [13] for details).

It remains for us to describe the isomorphism invariants of an arbitrary ZG-lattice M . Suppose that M is expressed as a direct sum of indecomposable ZG-lattices, chosen as above. The following are a full set of isomorphism invariants of M :

(a) A set of $4p + 1$ non-negative integers which determine the genus of M ; they are the multiplicities of the $4p + 1$ indecomposable $\hat{Z}G$-lattices as direct summands of $\hat{M}$.

(b) The R_1- and R_2-ideal classes associated with M . If M is expressed as an extension of an R_2-lattice N by a Λ_1-lattice L , then these ideal classes are just the Steinitz classes of L and N . In particular, when M is expressed as a direct sum of indecomposable ZG-lattices of the types arising when replacements (i)-(iv) are performed, then the R_1-ideal class of M is just the ideal class of the product of all of the R_1-ideals occurring in the summands of M , and likewise for the R_2-ideal class.

(c) There are two other possible invariants of M , each of which occurs only in certain special cases. These cases are specified by the genus invariants in (a), and depend also on whether $p \equiv 1 \pmod 4$ or not. One such possible invariant is an element of some group U_t of the type defined in the previous section. The other possible invariant, which can occur only when $p \equiv 1 \pmod 4$, is a certain quadratic character mod p (see [13] for details).

It may be worthwhile to point out why these last two possible invariants of M may fail to occur. This fact is a consequence of the Exchange and Absorption Formulas of Section 2. If $\underline{\underline{a}}_i$ denotes an ideal of R_1, and $\underline{\underline{b}}_i$ an ideal of R_2, then (with suitable choices for the exponent k and the element u) we obtain identities such as

$$\underline{\underline{a}}_1 \oplus \left\{\underline{\underline{a}}_2, \underline{\underline{b}}; \lambda^k u\right\} \cong \underline{\underline{a}}_1\underline{\underline{a}}_2 \oplus \left\{R_1, \underline{\underline{b}}; \lambda^k\right\},$$

$$E(\underline{\underline{a}}_1) \oplus \left\{Z \oplus E(\underline{\underline{a}}_2), \underline{\underline{b}}; 1 \oplus \lambda^k u\right\} \cong E(\underline{\underline{a}}_1\underline{\underline{a}}_2) \oplus \left\{Z \oplus \Lambda_1, \underline{\underline{b}}; 1 \oplus \lambda^k\right\},$$

and so on. These formulas can be used systematically to reduce the direct sum decomposition of M to some canonical form. Once this is done, the existence of the possible invariants described in (c) is proved by a detailed study of the orbits of $\mathrm{ext}(N, L)$ under the specifically-computed automorphism groups $\mathrm{aut}(N)$ and $\mathrm{aut}(L)$.

References

[1] С.Д. Берман, П.М. Гудивок [S.D. Berman, P.M. Gudivok], "Неразложимые представления конечных групп над кольцом целых p-адических чисел" [Indecomposable representations of finite groups over the ring of p-adic integers], *Izv. Akad. Nauk SSSR Ser. Mat.* **28** (1964), 875-910; English Transl., *Amer. Math. Soc. Transl.* (2) **50** (1966), 77-113.

[2] Charles W. Curtis, Irving Reiner, *Representation Theory of Finite Groups and Associative Algebras* (Pure and Applied Mathematics, **11**. Interscience [John Wiley & Sons], New York, London, 1962).

[3] Fritz-Erdmann Diederichsen, "Über die Ausreduktion ganzzahliger Gruppendarstellungen bei arithmetischen Äquivalenz", *Abh. Math. Sem. Univ. Hamburg* **13** (1940), 357-412.

[4] Steven Galovich, "The class group of a cyclic p-group", *J. Algebra* **30** (1974), 368-387.

[5] A. Heller and I. Reiner, "Representations of cyclic groups in rings of integers, I", *Ann. of Math.* (2) **76** (1962), 73-92.

[6] A. Heller and I. Reiner, "Representations of cyclic groups in rings of integers, II", *Ann. of Math.* (2) **77** (1963), 318-328.

[7] Alfredo Jones, "Groups with a finite number of indecomposable integral representations", *Mich. Math. J.* **10** (1963), 257-261.

[8] H. Jacobinski, "Genera and decompositions of lattices over orders", *Acta Math.* **121** (1968), 1-29.

[9] Michael A. Kervaire and M. Pavaman Murthy, "On the projective class group of cyclic groups of prime power order", *Comment. Math. Helv.* **52** (1977), 415-452.

[10] Irving Reiner, "A survey of integral representation theory", *Bull. Amer. Math. Soc.* **76** (1970), 159-227.

[11] I. Reiner, *Maximal Orders* (London Mathematical Society Monographs, 5. Academic Press [Harcourt Brace Jovanovich], London, New York, San Francisco, 1975).

[12] Irving Reiner, "Integral representations of cyclic groups of order p^2 ", *Proc. Amer. Math. Soc.* **58** (1976), 8-12. Erratum: *Proc. Amer. Math. Soc.* **63** (1977), 374.

[13] Irving Reiner, "Invariants of integral representations", *Pacific J. Math.* (to appear).

[14] Stephen V. Ullom, "Fine structure of class groups of cyclic p-groups", *J. Algebra* **49** (1977), 112-124.

[15] S. Ullom, "Class groups of cyclotomic fields and group rings", *J. London Math. Soc.* (to appear).

Department of Mathematics,
University of Illinois,
Urbana,
Illinois,
USA.

PROC. 18th SRI
CANBERRA 1978, 88-94.

20K20
(20K40)

ANNIHILATOR CLASSES OF TORSION-FREE ABELIAN GROUPS

Phillip Schultz

1. Introduction

Let F denote the class of non-zero torsion-free abelian groups. For any $A \in F$, the (*right*) *annihilator* of A, denoted $A^{\perp}$, is the class of all $B \in F$ for which $\hom(A, B) = 0$, and the *annihilator class* of A, denoted $[A]$, is the class of all $B \in F$ for which $A^{\perp} = B^{\perp}$. Annihilator classes arise in the study of torsion theories ([1], [4]); indeed, (L, R) is a torsion theory on a category of modules iff L is a collection of annihilator classes such that, for all $[A] \in L$, if $A^{\perp} \subseteq B^{\perp}$ then $[B] \in L$, while $R = \bigcap\{A^{\perp} : A \in L\}$.

In this paper, I show that the annihilator classes, under the operation induced by tensor product, and the relation induced by inclusion, form a small complete semi-lattice ordered monoid. This algebra has a "dense" subalgebra consisting of the annihilator classes of completely decomposable groups. In Section 2, I consider the elementary properties of the algebra of annihilator classes; Section 3 is concerned with the significance of the word "dense" above; finally in Section 4, I compute some annihilator classes.

2. The ordered monoid of annihilator classes

For $A, B \in F$, we define $[A] \leq [B]$ to mean $A^{\perp} \supseteq B^{\perp}$, and $[A].[B]$ to mean $[A \otimes B]$. The relation $\leq$ is clearly a well defined partial order; a routine calculation, using the usual adjoint relationship between hom and $\otimes$, shows that the multiplication $\cdot$ is also well defined and is compatible with the order. Let $\mathbb{A}$ be the collection of annihilator classes, and $\langle \mathbb{A}, \leq, \cdot \rangle$ the resulting algebra. By the customary abuse of notation, we use juxtaposition for multiplication, and $\mathbb{A}$ to denote the algebra as well as its universe.

In the following list of elementary facts about $\mathbb{A}$, the omission of a proof indicates that the proof is trivial.

(A) $\mathbb{A}$ is a small complete join semi-lattice. (The word "small" has the following significance: every genuine subset of $\mathbb{A}$ has a supremum, but $\mathbb{A}$ itself is a proper class in the sense of von Neumann-Bernays-Gödel set theory.)

Proof. Let $\mathbb{C}$ be a set of objects of $\mathbb{F}$, and $X \in \mathbb{F}$. Then $X \in (\bigoplus_{A \in \mathbb{C}} A)^{\perp}$ iff $X \in \bigcap_{A \in \mathbb{C}} A^{\perp}$, so $[\bigoplus_{A \in \mathbb{C}} A]$ is a supremum for the set $\{[A] : A \in \mathbb{C}\}$.

(B) $\mathbb{A}$ has minimum $[\mathbb{Q}]$ and maximum $[\mathbb{Z}]$.

(C) $\mathbb{A}$ is an ordered commutative monoid with zero $[\mathbb{Q}]$ and identity $[\mathbb{Z}]$.

(D) The completely decomposable groups form a subalgebra $\mathbb{B}$ of $\mathbb{A}$; the completely decomposable groups all of whose elements are of idempotent type form a subalgebra $\mathbb{C}$ of $\mathbb{B}$.

(E) $\mathbb{A}$ has as atoms the rank 1 p-local groups in $\mathbb{C}$.

(F) For all $A, B \in \mathbb{F}$, if $[A] \leq [B]$, then $\hom(B, A) \neq 0$.

From (F), we deduce that non-isomorphic rank 1 groups represent distinct classes in $\mathbb{A}$. Furthermore, a result of Shelah [6] that for any cardinal λ there is a rigid system with 2^{λ} elements, implies that $\mathbb{A}$ is a proper class.

(G) For all $A, B \in \mathbb{F}$, if A and B are quasi-isomorphic, then $[A] = [B]$.

(H) For all $A \in \mathbb{F}$, $[A]$ is closed under direct sums and extensions.

Proof. Closure under direct sums is easy.

Suppose $B \rightarrowtail C \twoheadrightarrow D$ is an exact sequence in $\mathbb{F}$ with $B, D \in [A]$. If $X \in A^{\perp}$ and $f \in \hom(C, X)$, then f restricted to B and the induced map on D are both zero, so $f = 0$. Conversely if $X \in C^{\perp}$, then $X \in D^{\perp} = A^{\perp}$.

(I) For all $A \in \mathbb{F}$, $\{B : [B] \leq [A]\}$ is closed under factors, direct sums, and extensions.

(J) If $A \rightarrowtail B \twoheadrightarrow C$ is a short exact sequence in $\mathbb{F}$, then $[C] \leq [B] \leq [A \oplus C] = \sup\{[A], [C]\}$.

3. The subalgebra of completely decomposable groups

Recall from (D), (A) and (F) above that $\mathbb{C} \subseteq \mathbb{B} \subseteq \mathbb{A}$ is a chain of subalgebras closed under suprema, and that for rank 1 groups A and B, $[A] \leq [B]$ iff there exists a monomorphism $B \rightarrowtail A$. The following result gives a similar criterion for completely decomposable groups.

LEMMA 1. *Let* $[A], [B] \in \mathcal{B}$. *Then* $[A] \leq [B]$ *iff for each rank* 1 *summand* A_i *of* A, *there is a rank* 1 *summand* B_j *of* B *such that* $B_j \rightarrowtail A_i$.

Proof (ONLY IF). Let $\{B_j : j \in J\}$ be a representative set of the rank 1 summands of B, so $[B] = \left[\bigoplus_{j\in J} B_j\right]$. Let A_i be any rank 1 summand of A. Since $A_i \notin A^{\perp}$, $\hom\left(\bigoplus_{j\in J} B_j, A_i\right) \cong \prod_{j\in J} \hom(B_j, A_i) \neq 0$, so there exists $j \in J$ with $\hom(B_j, A_i) \neq 0$. Hence there exists a monomorphism $B_j \rightarrowtail A_i$.

(IF). Let $X \in B^{\perp}$, so for each component B_j of B, $\hom(B_j, X) = 0$. Then for each component A_i of A, $\hom(A_i, X) = 0$ so $X \in A^{\perp}$.

COROLLARY *For* $[A]$, $[B] \in \mathcal{B}$, $[A] = [B]$ *iff each component of* A *is embedded in a component of* B, *and vice versa.*

We shall now show that every element of $\mathcal{A}$ has upper and lower bounds in $\mathcal{B}$ (other than the trivial ones $[\mathbb{Q}] \leq [A] \leq [\mathbb{Z}]$), and we shall consider in what circumstances these bounds are best possible.

PROPOSITION 1. *Let* $A \in \mathcal{F}$, *let* $\mathcal{D}$ *be the set of (isomorphism classes of) rank* 1 *torsion-free factor groups of* A, *and let* $\mathcal{G}$ *be the set of rank* 1 *pure subgroups of* A. *Then*

$$\Delta = \left[\bigoplus_{D\in\mathcal{D}} D\right] \leq [A] \leq \left[\bigoplus_{G\in\mathcal{G}} G\right] = \Gamma$$

and Δ *is the greatest lower bound for* $[A]$ *in* $\mathcal{B}$.

Proof. By (J), for each $D \in \mathcal{D}$, $[D] \leq [A]$, so $\Delta \leq [A]$. Let $M = \bigoplus_{E\in\mathcal{E}} E$ be a completely decomposable group such that $[M] \leq [A]$. For each $E \in \mathcal{E}$, $[E] \leq [A]$, so by (F), $\hom(A, E) \neq 0$, so A has a rank 1 factor $D \rightarrowtail E$. By Lemma 1, $[M] \leq \Delta$.

The inclusion of each $G \in \mathcal{G}$ into A induces a homomorphism $f : \bigoplus_{G\in\mathcal{G}} G \to A$ which is clearly surjective. By (J), $[A] \leq \Gamma$.

REMARKS. Later we shall see that Γ is not necessarily a least upper bound for $[A]$ in $\mathcal{B}$; in fact such least upper bounds may fail to exist. However, the following definition and proposition yield an upper bound for $[A]$ in $\mathcal{B}$ which is an improvement on Γ.

For any $A \in \mathcal{F}$, and any set T of types, define the *T-sequence* of A,

$T_1(A) \subseteq T_2(A) \subseteq \ldots \subseteq T_\nu(A) \subseteq \ldots$, as follows: $T_1(A)$ is the pure subgroup of A generated by $\{A[t] : t \in T\}$; if ν is an ordinal for which $T_\nu(A)$ has been defined,

$$T_{\nu+1}(A)/T_\nu(A) = T_1\left(A/T_\nu(A)\right) ;$$

if λ is a limit ordinal, $T_\lambda(A) = \bigcup_{\nu<\lambda} T_\nu(A)$.

Clearly the T-sequence of A is an increasing chain of pure subgroups of A which eventually terminates with $T_\nu(A) = T_\mu(A)$ for all $\mu > \nu$. A has *T-length* α if α is the least ordinal such that $T_\alpha(A) = T_\mu(A)$ for all $\mu > \alpha$. We say T *fills* A if $T_\alpha(A) = A$. For example, if T contains the typeset of A , then A has T length 1 and T fills A . If $t' \leq t$ are types and $\{t\}$ fills A , then also $\{t'\}$ fills A .

PROPOSITION 2. *Let* $A \in F$ *and let* T *be a set of types. For each* $t \in T$, *let* $B(t)$ *be a rank* 1 *group of type* t , *and let* $B = \bigoplus_{t \in T} B(t)$. *Then* $[A] \leq [B]$ *iff* T *fills* A .

Proof (ONLY IF). Let A have T-length α , so $A/T_\alpha(A)$ has no element of type greater than or equal to t for all $t \in T$, so $\hom\left(B, A/T_\alpha(A)\right) = 0$. But $\left[A/T_\alpha(A)\right] \leq [A] \leq [B]$, so by (F), $A/T_\alpha(A) = 0$.

(IF). If $X \in B^\perp$, then $X[t] = 0$ for all $t \in T$, so $\left[T_1(A)\right]$ and each $\left[T_{\nu+1}(A)/T_\nu(A)\right] \leq [B]$. By (I), if $\left[T_\nu(A)\right] \leq [B]$, then $\left[T_{\nu+1}(A)\right] \leq [B]$. If λ is a limit ordinal, and $\left[T_\nu(A)\right] \leq [B]$ for all $\nu < \lambda$, then $\left[T_\lambda(A)\right] \leq [B]$. Hence by induction, $[A] = \left[T_\alpha(A)\right] \leq [B]$, where A has T-length α .

EXAMPLE ([5]). Let (T, T') be a countable set of pairs (t, t') of types satisfying:

(1) there exists t such that for all $\left(t_1, t_1'\right), \left(t_2, t_2'\right) \in (T, T')$,

$$t_1 \wedge t_2 = t ;$$

(2) there exists s such that for all $(t, t') \in (T, T')$,

$$t.t' = s ;$$

(3) for all $(t, t') \in (T, T')$, t and t' are incomparable;

(4) for all $\left(t_1, t_1'\right) \neq \left(t_2, t_2'\right) \in (T, T')$, $t_1 < t_2'$.

It is proved in [5] that there are infinitely many strongly indecomposable rank

2 groups A having T as typeset and such that, for each $(t, t') \in (T, T')$, $A[t]$ has a rank 1 and $A/A[t]$ has type t'.

For each $(t, t') \in (T, T')$, let $B(t)$ be a rank 1 group of type t, $B(t')$ a rank 1 group of type t'. Then for all $(t, t') \in (T, T')$,

$$[A] \leq [B(t) \oplus B(t')] .$$

Moreover, if C is any completely decomposable group such that

$$[A] \leq [C] ,$$

then by Proposition 2, C has a rank 1 component D of type r such that $A[r] \neq 0$. Hence for some $(t, t') \in (T, T')$, $D \rightarrowtail B(t)$, so $[B(t)] \leq [D] \leq [C]$.

By (J), $[B(t')] \leq [A] \leq [C]$, so $[B(t) \oplus B(t')] \leq [C]$. We have shown that if $[A]$ has any minimal upper bound in $\mathcal{B}$, it must be some $[B(t) \oplus B(t')]$. On the other hand, by (4),

$$[B(t_1) \oplus B(t_1')] \leq [B(t_2) \oplus B(t_2')] \quad \text{iff} \quad B(t_2') \rightarrowtail B(t_1') .$$

For suitable choices of (T, T'), this may or may not occur. $B(t') \rightarrowtail B(t_1')$ for all (t, t') iff $[B(t_1) \oplus B(t_1')]$ is a least upper bound for $[A]$ in $\mathcal{B}$; $B(t_1') \rightarrowtail B(t')$ for no $(t, t') \neq (t_1, t_1')$ iff $[B(t_1) \oplus B(t_1')]$ is a minimal upper bound for $[A]$ in $\mathcal{B}$.

4. Equivalence in $\mathcal{B}$

In general, it seems to be hard to compute $[A]$ for arbitrary $A \in \mathcal{F}$. However, in view of the density theorems of the previous section, it would be useful to know which groups are equivalent to completely decomposable ones. The following result follows immediately from Propositions 1 and 2.

LEMMA 2. *Let* $A \in \mathcal{F}$, *and let* B *be a rank* 1 *group of type* t. *Then* $[A] = [B]$ *iff* B *is a factor group of* A *and* $\{t\}$ *fills* A.

EXAMPLE. Let t be a nil type, and suppose a rank n group C such that $[C] = [B]$ has been constructed. There is an exact sequence

$$\mathbb{Z} \rightarrowtail B \twoheadrightarrow T , \text{ where } T = \bigoplus_{p \in S} \mathbb{Z}(p^{k(p)}) ,$$

$0 < k(p) \leq \infty$ for all $p \in S$, and for infinitely many $p \in S$, $k(p) < \infty$. Hence there is an exact sequence

$$\hom(\mathbb{Z}, C) \to \operatorname{ext}(T, C) \twoheadrightarrow \operatorname{ext}(B, C) ,$$

in which $\hom(\mathbb{Z}, C) \cong C$, and by [2, p. 224], $\operatorname{ext}(T, C) \cong \hom(T, D/C)$, where D is the divisible hull of C. Since B is a rank 1 factor of C which has finite type at infinitely many primes $p \in S$, D/C has nonzero p-component at infinitely many $p \in S$, so $\hom(T, D/C)$ has a subgroup isomorphic to $\prod_{p \in U} \hom(\mathbb{Z}(p^{k(p)}), \mathbb{Z}(p^{\infty}))$ where U is an infinite subset of S .

Hence $\mathrm{ext}(T, C)$ is uncountable, and since C is countable, $\mathrm{ext}(B, C)$ is uncountable. Thus there are uncountably many non-congruent extensions of C by B , and hence uncountably many non-isomorphic ones. By (H) of Section 2, each such extension is equivalent to B , so for each positive integer $n > 1$, there are uncountably many rank n groups C with $[C] = [B]$.

For idempotent types, the characterization is a little easier.

PROPOSITION 3. *Let* $A \in F$ *, and let* B *be a rank* 1 *group of idempotent type* t *. Then* $[A] = [B]$ *iff* $A \cong C \oplus B$ *for some* B*-module* C .

Proof (ONLY IF). By Proposition 2, $\{t\}$ fills A , so $A[t] \neq 0$. But for idempotent t , $(A/A[t])[t] = 0$ [3] , so $A = A[t]$ and hence A is a B-module. By Proposition 1, there is an exact sequence of abelian groups

$$(*) \qquad C \rightarrowtail A \twoheadrightarrow B .$$

But since C is pure in A , (*) is an exact sequence of B-modules, and therefore splits.

(IF). By (J) of Section 2, $[B] \leq [A] \leq [B \oplus C]$, but $[B]$ is maximal among the classes of B-modules in A , so $[A] = [B]$.

5. Acknowledgements

This paper is part of a more general study of the algebra of annihilator classes in an abelian category. In this larger context, many of the ideas are due to Rod Bowshell.

I acknowledge the excellent facilities and inspiring atmosphere of the 1978 Summer Research Institute of the Australian Mathematical Society at the Australian National University where this paper was written.

References

[1] Carl Faith, *Algebra: Rings, Modules and Categories* I (Springer-Verlag, Berlin, Heidelberg, New York, 1973).

[2] László Fuchs, *Infinite Abelian Groups*, Vol. 1 (Pure and Applied Mathematics, **36**. Academic Press, New York, London, 1970).

[3] B.J. Gardner, "A note on types", *Bull. Austral. Math. Soc.* 2 (1970), 275-276.

[4] Joachim Lambek, *Torsion Theories, Additive Semantics, and Rings of Quotients* (Lecture Notes in Mathematics, **177**. Springer-Verlag, Berlin, Heidelberg, New York, 1971).

[5] Phillip Schultz, "The typeset and cotypeset of a rank 2 abelian group", *Pacific J. Math.* (to appear).

[6] Saharon Shelah, "Infinite abelian groups, Whitehead problem and some constructions", *Israel J. Math.* 18 (1974), 243-256.

Department of Mathematics,
University of Western Australia,
Nedlands,
Western Australia.

PROC. 18th SRI
CANBERRA 1978, 95-107.

13-01, 15-01
(13B25, 15A36)

THE ROLE OF ALGORITHMS IN THE TEACHING OF ALGEBRA

Charles C. Sims

For the last several years I have been working in the field of group-theoretic algorithms and my second lecture [11] at this Summer Research Institute will be devoted to a survey of some of the more important algorithms which have been developed to solve problems in group theory. Other speakers have also provided evidence that the desire to obtain constructive solutions to specific mathematical problems motivates a great deal of research activity in algebra. For example, Professor Baker discussed some results which show that algorithms exist for solving certain types of Diophantine equations which have been studied for several hundred years. Also, Professor Rabin in his video-taped lectures described some very efficient algorithms for solving problems about polynomials over finite fields.

In this talk I want to express my concern that although the subject of algebraic algorithms is a significant part of research in algebra we have too often failed to show our students that we consider it important to have algorithmic solutions for classes of algebraic problems. Even in situations where a problem and an efficient algorithm for its solution are accessible to beginning algebra students we have neglected to take the time to let students see the algorithm and work with it enough to gain a good insight into its operation.

At this point it would be useful to have before us an example of an algebraic algorithm. The algorithm which makes the Fundamental Theorem of Finitely Generated Abelian Groups constructive provides a good illustration. Let G be an abelian group generated by n elements. There is a homomorphism f from the free abelian group $\mathbb{Z}^n$ onto G. The kernel M of f is finitely generated and so there is an m-by-n integer matrix A such that the rows of A generate M. Let us write $M = S(A)$. There is a procedure for reducing A by means of elementary row and column operations to a matrix D in Smith normal form, that is,

$$D = \begin{bmatrix} d_1 & & & & \\ & d_2 & & 0 & \\ & & \ddots & & \\ 0 & & & d_r & \\ & & & & 0 \end{bmatrix}$$

with each d_i a positive integer such that d_i divides d_{i+1} for $1 \quad i < r$. This procedure, together with the fact that

$$G \cong \mathbb{Z}^n/S(A) \cong \mathbb{Z}^n/S(D) \cong \mathbb{Z}_{d_1} \times \ldots \times \mathbb{Z}_{d_r} \times \mathbb{Z}^{n-r},$$

makes it possible to express G as a direct sum of cyclic groups.

It is probably not right to think of reduction to Smith normal form as a group-theoretic algorithm at all, but rather as an algorithm related to modules over Euclidean domains, or even, with a generalization of the operations performed, to modules over principal ideal domains. For example, let K be a field and let B be an n-by-n matrix with entries in K. If we consider $B - xI$ to be a matrix with entries in $K[x]$, then we can reduce $B - xI$ by row and column operations to a matrix C with

$$C = \begin{bmatrix} f_1 & & & 0 \\ & f_2 & & \\ & & \ddots & \\ 0 & & & f_n \end{bmatrix},$$

where each f_i is a monic polynomial and f_i divides f_{i+1} for $1 \le i < n$. The rational canonical form for B is the direct sum of the companion matrices for those of the f_i which have positive degree. Thus the question of similarity of matrices over a field has an algorithmic solution.

One of the reasons that I became interested in the emphasis placed on algorithms in the teaching of algebra is the performance of graduate students on qualifying examinations in which I have participated. In my experience too many students master the definitions and theorems of algebra without developing the ability to apply the ideas involved to work out specific examples. In a word, students can not compute! Typically students can state the Fundamental Theorem of

Finitely Generated Abelian Groups but when given the group $G = \mathbb{Z}^2/S(A)$, where

$$A = \begin{bmatrix} -9 & -6 \\ 12 & 6 \end{bmatrix},$$

they are unable to deduce that $G \cong \mathbb{Z}_3 \times \mathbb{Z}_6$.

To see whether this was a purely local phenomenon, peculiar to my own university, I decided to look at several well known algebra texts. I selected four problems of a computational nature and attempted to determine whether a student could find in any of the texts a statement about the existence of algorithmic solutions for these problems. The problems were:

1. Given an m-by-n integer matrix A , compute the orders of the cyclic direct factors of the abelian group $\mathbb{Z}^n/S(A)$.
2. Given f in $\mathbb{Z}[x]$, factor f into irreducible factors.
3. Given f irreducible in $\mathbb{Q}[x]$, compute the Galois group of f over $\mathbb{Q}$.
4. Given $f_1, \ldots, f_r$ and g in $\mathbb{Z}[x]$, decide whether g is in the ideal of $\mathbb{Z}[x]$ generated by the f_i .

Each of these problems has an algorithmic solution and I shall briefly discuss the solutions in a moment.

The five texts which I consulted, listed in the order of their initial publication, were the books by Weber [14], van der Waerden [13], Jacobson [5], Lang [7] and Mac Lane and Birkhoff [8]. Before giving the results of my survey, I want to emphasize that I did not attempt to read every word of each of these books. I simply looked at those sections which I thought a student might reasonably consult in order to find out something about the problems stated above. Thus it is possible that I have overlooked some reference to these problems. If this has happened, I would like to be informed about it and I offer my apologies in advance for any such omission.

The following table summarizes the results I obtained. The books are referred to in the order given above by the abbreviations W, vdW, J, L and MB.

Problem	W	vdW	J	L	MB
1	*N*	*N*	*Y*	*N*	*Y*
2	*Y*	*Y*	*N*	*N*	*N*
3	*N*	*Y*	*N*	*N*	-
4	*N*	*N*	*N*	*N*	*N*

An entry of "Y" indicates that at least the statement that the problem in question possesses a finite solution was found, while an "*N*" indicates the absence of any such statement. The dash for the third problem in the column headed MB is to point out that Mac Lane and Birkhoff do not discuss Galois theory and so of course no discussion of problem 3 is included.

The algorithm for reducing an integer matrix to Smith normal form is essentially a two-dimensional version of the Euclidean algorithm for computing greatest common divisors. It and its generalization to matrices over a principal ideal domain may be found in [8].

There are several ways of showing that a polynomial f in $\mathbb{Z}[x]$ can effectively be factored into irreducible factors. Let f have degree n and let g be a factor of f. We may assume the degree of g is between 1 and $m = [n/2]$. Select $m+1$ distinct integers $a_0, \ldots, a_m$. If $f(a_i) = 0$ for some i, then $x - a_i$ is a factor of f. Thus we may assume $f(a_i) \neq 0$ for all i. Then $g(a_i)$ must be one of the finitely many divisors of the integer $f(a_i)$. Given integers $b_0, \ldots, b_m$ such that b_i divides $f(a_i)$, we can interpolate a unique polynomial g with rational coefficients of degree at most m such that $g(a_i) = b_i$. If g does not have integer coefficients, then we can go on to another choice of the b_i. Otherwise we must check to see whether or not g divides f. Another approach to factoring f depends on the observation that we can bound the absolute value of the roots of f in the field of complex numbers and in turn bound the size of the coefficients of any factor of f. Neither of these methods turns out to be very efficient in practice and better methods which involve factoring f modulo p for various primes p have been developed. The survey by Zimmer [15] has additional references.

In [13] it is shown that the computation of the Galois group of an irreducible polynomial f in $\mathbb{Q}[x]$ is a finite problem. However, the procedure described there is not practical for polynomials of even moderately large degree. We can multiply f by a suitable integer and then make a linear change of variable in such a way that we obtain a monic polynomial g in $\mathbb{Z}[x]$ with the same Galois group. Computing the Galois group of g considered as an element of $\mathbb{Z}_p[x]$ for several primes p provides information about the cycle types of elements in the Galois group of g over $\mathbb{Q}$ in its permutation representation on the roots of g. Further information on this problem can also be found in [15] and [12].

The solution of the fourth problem given above is more difficult to find in the literature. Hilbert's Basis Theorem appeared in 1890 [4]. This theorem implies that any ideal in $R = \mathbb{Z}[x_1, \ldots, x_n]$ or in $S = \mathbb{Q}[x_1, \ldots, x_n]$ is finitely generated. In [3] it is proved that given elements $f_1, \ldots, f_r$ and g of S we can decide whether g is in the ideal of S generated by $f_1, \ldots, f_r$. A remark on pages 24-25 of [15] indicates that computer programs have been written to handle the corresponding problem for R when $n = 1$. A solution for this problem is given in the appendix to this paper. A similar algorithm for $n > 1$ can be formulated.

It is my opinion that students who have had a graduate level course in algebra and do not know that the four problems given above have algorithmic solutions have missed something important.

It is worth remarking that even van der Waerden, who among the authors of the five texts involved in the survey appears the most interested in algebraic algorithms, takes a short cut in the proof of the Fundamental Theorem of Finitely Generated Abelian Groups which makes his proof not constructive. Roughly speaking, when he is attempting to prove that any integer matrix A is equivalent to a matrix in Smith normal form, he asks that we consider, among all matrices equivalent to A, a matrix B in which a nonzero entry of smallest absolute value occurs. It takes only a few more lines to show how to produce such a B explicitly in a finite number of steps.

Another example of what I would consider a lack of proper concern for algorithmic questions in algebra can be found in [7]. On page 128, we find the statement: "It is usually not too easy to decide when a given polynomial (say in one variable) is irreducible. For instance, the polynomial $X^4 + 4$ is *reducible* over the rational numbers, because

$$X^4 + 4 = \left(X^2-2X+2\right)\left(X^2+2X+2\right) ."$$

The student is left with the impression that had someone not stumbled across the factorization of $X^4 + 4$ we might never have known that $X^4 + 4$ was reducible. In the same space one can describe state of affairs much more accurately. Polynomials with rational coefficients can be factored into irreducible factors but the algorithms we know involve a nontrivial amount of computation.

Let me now describe my favorite algebraic algorithm, which is so simple and elegant that I think it should be shown to all graduate students in algebra. The algorithm is due to Berlekamp [1] although the formulation given here is different from the one given by him.

Let A be a commutative algebra of finite dimension n over the finite field $K = \mathrm{GF}(q)$ with q elements. The map $T : A \to A$ given by $T(a) = a^q$ is a linear transformation of A.

THEOREM (Berlekamp). *The algebra A is a field if and only if both of the following conditions hold:*

(a) T *is nonsingular;*

(b) $T - I$ *has rank* $n - 1$ *, where* I *is the identity transformation.*

Given the structure constants for A relative to some basis, the matrix of T with respect to the same basis can be found and conditions *(a)* and *(b)* can be easily checked. Berlekamp was interested in the case where A is given as the quotient algebra $K[x]/M$, where M is the ideal generated by a given polynomial f. The theorem gives an efficient irreducibility test for f.

Although it is important to show students examples of interesting algebraic algorithms, students must not be left with the impression that every computational problem in algebra has an algorithmic solution and that it is simply a matter of being clever enough to find the algorithm. Students need to be told that there are problems for which it can be proved that there is no algorithm for producing a solution. Some of the first such problems found are connected with finitely presented groups.

Let X be a set and let F be the free group generated by X. The elements of F are equivalence classes of words, where by a *word* we mean a finite sequence of elements from the set $X \times \{1, -1\}$. It is traditional to write (x, ε) as x^{ε}. Two words U and V are equivalent if one can be obtained from the other by inserting and deleting consecutive terms of the form $x^{\varepsilon}, x^{-\varepsilon}$. The equivalence class containing the word U will be denoted $[U]$.

Let R be a set of words. The subgroup $N = N(\mathsf{R})$ of F generated by all conjugates of the elements $[R]$ with R in R is normal in F. We denote the quotient group F/N by $\langle X|\mathsf{R}\rangle$. The pair X, R is a presentation for a group G if $G \cong \langle X|\mathsf{R}\rangle$. The presentation is finite if both X and R are finite sets. For example, if we take X to be the two-element set $\{x, y\}$ and take

$$\mathsf{R} = \{x^2, y^3, (xy)^5\} ,$$

then $\langle X|\mathsf{R}\rangle$ is isomorphic to the alternating group A_5 and so X, R is a finite presentation for A_5.

Suppose X, R is a finite presentation for a group G. We may think of G as being $\langle X|\mathsf{R}\rangle$. It is natural to ask the following question: Given a word U, can we decide whether U represents the identity in G, or equivalently, whether $[U]$ is in $N(\mathsf{R})$? This is referred to as the *word problem* for the presentation X, R. Some twenty years ago Novikov [9] and Boone [2] showed that there are finite presentations for which the word problem can not be solved by any algorithm. Many other similar results followed. We can not in general decide when two words represent conjugate elements of G. Neither can we decide whether G is finite nor whether G has more than one element. A more detailed discussion of computational problems related to finitely presented groups can be found in [11].

Professor Kaplansky discussed another negative result. Hilbert's 10th problem has no solution. There is no algorithm for answering the following question: Given a polynomial $f(x_1, \ldots, x_n)$ in $\mathbb{Z}[x_1, \ldots, x_n]$, do there exist integers $a_1, \ldots, a_n$ such that $f(a_1, \ldots, a_n) = 0$?

It is interesting to note that the result of Novikov and Boone has not deterred group theorists from writing down presentations of groups and attempting to study the groups so defined. Similarly, the fact that there is no general algorithm for solving Diophantine equations has not stopped people from writing down Diophantine equations and trying to solve them. I consider this evidence of the innate optimism of mathematicians. With both the word problem and the problem of solving Diophantine equations we have the same situation. The general problem is known to have no algorithmic solution while algorithms for solving many special cases are known. I find it exciting to contemplate how narrow the gulf between these two extremes can be made. It should be observed that the methods used to solve the special cases involve traditional algebraic techniques while the methods used to show the general case has no algorithmic solution are on the whole foreign to most algebraists. Anyone who is going to work on both sides of this gulf must be trained in traditional algebra and in formal logic. I consider this a strong argument for exposing algebra students to more formal logic than is customary.

Suppose for a moment that we agree that a discussion of algebraic algorithms is an important part of the teaching of algebra. There is still the very real problem of fitting this material into an already crowded syllabus. Every text for a graduate level basic algebra course contains more material than can possibly be covered by a lecturer in one year. How can we add even more material with which we expect the well-educated algebraist to be familiar? My answer is that many of the algorithms can be covered best in an introductory undergraduate course. Quite often graduate courses repeat a great deal of material given in undergraduate courses. To me it would be more efficient to omit a few of the deeper theorems often proved in undergraduate courses and use the time for a discussion of a representative selection of algebraic algorithms. Having the students experiment with algorithms on concrete problems can provide a better intuitive grasp of the concepts of groups, rings and fields than can piling theorem upon theorem. As an example, the Fundamental Theorem of Galois Theory is one topic which in my view does not need to be covered at the undergraduate level. In its place one might cover some of the algorithms discussed by Professor Rabin for factoring polynomials and finding roots of polynomials over finite fields.

It is an unfortunate fact that most algebraic algorithms require a fair amount of computation even when applied to simple examples, too much computation for an undergraduate just learning the basic ideas to carry out by hand without making many errors. Thus some kind of computer assistance should be provided. There is no generally accepted method for accomplishing this. One approach is to teach the students a programming language and ask them to write their own programs implementing various standard algorithms. A very different approach is to provide the students with "canned programs" which the students can use without any knowledge of computer programming. The first approach wastes too much of the students' time in purely programming details while the second does not give enough exposure to the actual operation of the algorithms. Some balance between the two seems desirable.

An example of an algorithm which can be nicely treated early in an introductory algebra course is the primality test discussed on pages 347-348 of [6]. The only prerequisites are a little group theory, Lagrange's Theorem and its corollary that if x is an element of the finite group G, then $x^{|G|} = 1$, together with the definition of the ring $\mathbb{Z}_n$ of integers modulo n and the result that the group U_n of units in $\mathbb{Z}_n$ consists of those congruence classes containing integers relatively prime to n. Thus the students can see that n is a prime if and only if $|U_n| = n - 1$. This particular primality test assumes that the prime factors of $n - 1$ are known and proceeds as follows:

1. Select an element x of $\mathbb{Z}_n$ with $x \neq 0, 1$.
2. Compute x^{n-1}. (This can be done with $O(\log n)$ multiplications modulo n.)
3. If $x^{n-1} \neq 1$, then n is not a prime.
4. If $x^{n-1} = 1$, then compute the order m of x in U_n using the fact that if $x^m = 1$, then either m is the order of x or $x^{m/p} = 1$ for some prime factor p of m.
5. If $m = n - 1$, then n is prime.

If the order m of x is not $n - 1$, then we repeat the procedure with other values of x and compute the least common multiple of their orders. It can happen that $|U_n|$ is a proper divisor of $n - 1$. However, in this case a randomly chosen x in $\mathbb{Z}_n$ will be a nonunit with probability greater than $\frac{1}{2}$. If x is a nonunit, then $x^{n-1} \neq 1$. The probabilistic primality test of Rabin [10] is a very powerful algorithm which can be described to undergraduates but the theory involved is probably too complicated to be discussed in full.

The correctness of opinions concerning the way mathematics should be taught can not be demonstrated in the same way that the correctness of a mathematical proof can be decided. I will consider this talk a success if the next time you pick up your favorite algebra text you look at it from a new point of view and if the next time you plan a syllabus for an introductory algebra course you think through carefully what algorithms you will present to your students.

APPENDIX

This appendix contains my solution to the fourth problem stated above. It is based on the idea of constructing for the ideal I generated by a given finite subset S of $\mathbb{Z}[x]$ a basis of the type exhibited in the proof of the Hilbert Basis Theorem as it is proved, for example, in [7]. If f is a nonzero element of $\mathbb{Z}[x]$, then $\deg(f)$ will denote the degree of f and $l(f)$ will denote the leading coefficient of f. For the purposes of this discussion, let us say that a finite subset T of $\mathbb{Z}[x]$ is *uniform* if the following conditions hold:

(a) T does not contain 0;

(b) T does not contain two different elements of the same degree;

(c) if T contains elements of degrees m and n with $m < n$, then for each integer i with $m < i < n$ there is an element of T of degree i.

LEMMA 1. *If S is a finite subset of $\mathbb{Z}[x]$, then there is a uniform subset T of $\mathbb{Z}[x]$ such that S and T generate the same ideal in $\mathbb{Z}[x]$.*

Proof. I shall describe a procedure Q for constructing one such subset T from S. The verification that Q performs correctly is left to the reader. The steps of Q are:

1. Set $T = S - \{0\}$.
2. If T does not have two different elements of the same degree, then go to Step 4.
3. Choose two different elements f and g of T of the same degree n with n as large as possible. Let $a = l(f)$ and $b = l(g)$. We may assume $|a| \le |b|$. Find integers q and r such that $b = qa + r$ and $0 \le r < |a|$. If $g = qf$, then delete g from T. Otherwise, replace g by $g - qf$. Go to Step 2.
4. If T contains an element f of degree m, an element of degree n with $n > m$ but no element of degree $m + 1$, then add xf to T and repeat this step.
5. Stop.

If T is a nonempty uniform subset of $\mathbb{Z}[x]$, then we can write T uniquely as $\{g_m, g_{m+1}, \ldots, g_M\}$, where m and M are the minimum and maximum degrees of elements of T, respectively, and g_i has degree i, $m \le i \le M$. Define $I(T)$ to be the set of all polynomials of the form

$$fg_M + \sum_{k=m}^{M-1} c_k g_k ,$$

where f is in $\mathbb{Z}[x]$ and the c_k are integers. If $T = \emptyset$, then define $I(T) = \{0\}$. Clearly $I(T)$ is always an additive subgroup of $\mathbb{Z}[x]$. Given a uniform subset T of $\mathbb{Z}[x]$ and an element h of $\mathbb{Z}[x]$, the following procedure R constructs an element $u = R(T, h)$ of the coset $I(T) + h$.

1. Set $u = h$. If $T = \emptyset$, then stop. Otherwise let $T = \{g_m, \ldots, g_M\}$ as above.
2. If $u = 0$, then stop. Otherwise, let $n = \deg(u)$.
3. If $n < m$, then stop.
4. If $n \le M$, then set $g = g_n$. Otherwise, set $g = x^{n-M} g_M$.
5. Let $a = l(g)$ and let b be the coefficient of x^n in u. Find integers q and r such that $b = qa + r$ and $0 \le r < |a|$. Replace u by $u - qg$, set $n = n - 1$ and go to Step 3.

The following lemmas are easily proved.

LEMMA 2. *Let* h_1 *and* h_2 *be in* $\mathbb{Z}[x]$ *and let* T *be a uniform subset of* $\mathbb{Z}[x]$. *Then* $I(T) + h_1 = I(T) + h_2$ *if and only if* $R(T, h_1) = R(T, h_2)$. *In particular,* h_1 *is in* $I(T)$ *if and only if* $R(T, h_1) = 0$.

LEMMA 3. *If* T *is a uniform subset of* $\mathbb{Z}[x]$, *then* $I(T)$ *is an ideal if and only if either* $T = \emptyset$ *or* $T = \{g_m, \ldots, g_M\}$ *as above and for all* i *with* $m \le i < M$ *the polynomial* xg_i *is in* $I(T)$.

Now we can describe an algorithm P which, given a finite subset S of $\mathbb{Z}[x]$, constructs a uniform subset T such that $I = I(T)$ is the ideal generated by S. Once we have T, Lemma 2 shows us how to decide whether a given polynomial h is in I. The steps of P are:

1. Set $T = S$.
2. Replace T by the result of applying algorithm Q to T.
3. If $T = \emptyset$ or if $R(T, xg) = 0$ for all g in T with less than the maximum degree, then stop.
4. Choose g in T of less than the maximum degree and such that $u = R(T,xg) \ne 0$. Set $T = T \cup \{u\}$ and go to Step 2.

By Lemma 3, if P terminates, then T has the desired properties. If $S = \emptyset$ or $S = \{0\}$, then P stops the first time Step 3 is reached. Suppose S contains a nonzero polynomial. Then T will never be empty. Suppose after some execution of Step 2 we have $T = \{g_m, \ldots, g_M\}$, where g_i has degree i , and after the next execution of Step 2 we have $T = \{h_n, \ldots, h_N\}$, where h_j has degree j . Then it is not too hard to show that $M = N$ and either $n < m$ or $n = m$ and

$$\sum_{i=m}^{M} |l(g_i)| > \sum_{i=m}^{M} |l(h_i)| .$$

This proves that Step 2 can be executed only finitely many times.

Let us consider an example. Suppose $S = \{f_1, f_2, f_3\}$, where

$$f_1 = 9 - 3x ,$$

$$f_2 = 3 - 9x + 6x^2 ,$$

$$f_3 = 1 + 11x + 10x^2 .$$

Define

$$f_4 = f_3 - f_2 = -2 + 20x + 4x^2 ,$$

$$f_5 = f_2 - f_4 = 5 - 29x + 2x^2 ,$$

$$f_6 = f_4 - 2f_5 = -12 + 78x ,$$

$$f_7 = f_6 + 26f_1 = 222 .$$

Applying algorithm Q to S gives $T = \{f_7, f_1, f_5\}$. Now $R(T, xf_7) = 0$ but $R(T, xf_1) = f_8 = 79 + 2x + x^2$. Applying Q to $\{f_7, f_1, f_5, f_8\}$, we introduce

$$f_9 = f_5 - 2f_8 = -153 - 33x ,$$

$$f_{10} = f_9 - 11f_1 = -252 ,$$

$$f_{11} = f_{10} + 2f_7 = 192 ,$$

$$f_{12} = f_7 - f_{11} = 30 ,$$

$$f_{13} = f_{11} - 6f_{12} = 12 ,$$

$$f_{14} = f_{12} - 2f_{13} = 6 ,$$

and obtain $T = \{f_{14}, f_1, f_8\}$. For $g = f_{14}$ and $g = f_1$ we have $R(T, xg) = 0$ and so the ideal I generated by S is $I(T)$.

Suppose we want to decide whether $h = 3 + x - x^2 + x^3$ is in I . We compute $R(T, h)$ and find that the answer is 0 . Using the quotients calculated in Step 5 of R , we see that

$$h = (x-3)f_8 + 24f_1 + 4f_{14} .$$

Moreover, we have the information necessary to express h in terms of f_1, f_2 and f_3 if we wish.

References

[1] E.R. Berlekamp, "Factoring polynomials over finite fields", *Bell System Tech. J.* **46** (1967), 1853-1859.

[2] William W. Boone, "Certain simple, unsolvable problems of group theory", *K. Nederl. Akad. Wetensch. Proc. Ser. A* [*Indag. Math.*]
I **57** [**16**] (1954), 231-237;
II **57** [**16**] (1954), 492-497;
III **58** [**17**] (1955), 252-256;
IV **58** [**17**] (1955), 571-577;
V **60** [**19**] (1957), 22-27;
VI **60** [**19**] (1957), 227-232.

[3] Grete Herman, "Die Frage der endlichen vielen Schritte in der Theorie der Polynomideale", *Math. Ann.* **95** (1926), 736-788.

[4] David Hilbert, "Ueber die Theorie der algebraischen Formen", *Math. Ann.* **36** (1890), 473-534.

[5] Nathan Jacobson, *Lectures in Abstract Algebra*. I: *Basic Concepts;* II: *Linear Algebra;* III: *Theory of Fields and Galois Theory* (The University Series in Higher Mathematics. Van Nostrand, Princeton, New Jersey; Toronto; New York; London; 1951, 1953, 1964. Reprinted: Graduate Texts in Mathematics, **30**, **31**, **32**. Springer-Verlag, New York, Heidelberg, Berlin, 1975.

[6] Donald E. Knuth, *The Art of Computer Programming*. Vol. 2: *Seminumerical Algorithms* (Addison-Wesley Series in Computer Science and Information Processing. Addison-Wesley, Reading, Mass.; Menlo Park, California; London; Don Mills, Ontario; 1969).

[7] Serge Lang, *Algebra* (Addison-Wesley Series in Mathematics. Addison-Wesley, Reading, Mass., 1965).

[8] Saunders Mac Lane, Garrett Birkhoff, *Algebra* (Macmillan, New York; Collier-Macmillan, London, 1967).

[9] П.С. Новиков [P.S. Novikov], "Об алгоритмической неразрешимости проблемы тождества слов в теории групп" [On the algorithmic insolvability of the word problem in group theory], *Trudy Mat. Inst. Steklov.* **44** (1955), 3-143; *Amer. Math. Soc. Transl.* (2) **9** (1958), 1-122.

[10] Michael O. Rabin, *Probabilistic Algorithms, Algorithms and Complexity* (Proc. Sympos. Carnegie-Mellon University, Pittsburgh, 1976, 21-39. Academic Press [Harcourt Brace Jovanovich], New York, London, 1976).

[11] Charles C. Sims, "Some group-theoretic algorithms", these proceedings, 108-124.

[12] Richard P. Standuhar, "The determination of Galois groups", *Math. Comp.* **27** (1973), 981-996.

[13] B.L. van der Waerden, *Moderne Algebra*, I, II (Unter Benutzung von Vorlesungen von E. Artin und E. Noether. Die Grundlehren der Mathematischen Wissenschaften, **23, 24**. J. Springer, Berlin, 1930, 1931. English editions: translated by Fred Blum and John R. Schulenberger. Frederick Unger, New York, 1970).

[14] Heinrich Weber, *Lehrbuch der Algebra*, I, II, III (F. Vieweg, Braunschweig, 1895, 1896, 1891. 2nd Editions: F. Vieweg, Braunschweig, 1898, 1899, 1908. Reprinted: Chelsea, New York, 1961). Original Title, 1891: *Elliptische Functionen und Algebraische Zahlen.*

[15] Horst G. Zimmer, *Computational Problems, Methods, and Results in Algebraic Number Theory* (Lecture Notes in Mathematics, **262**. Springer-Verlag, Berlin, Heidelberg, New York, 1972).

Department of Mathematics,
Rutgers University,
New Brunswick,
New Jersey,
USA.

PROC. 18th SRI
CANBERRA 1978, 108-124.

SOME GROUP-THEORETIC ALGORITHMS

Charles C. Sims

Group theorists have been attempting to construct algorithms for working with groups, both finite and infinite, for many years. It turns out that in order to solve a problem about a given finite group it is often necessary, or at least useful, to study one or more infinite groups in the process. One of the purposes of this talk is to give an example of an algorithm designed to provide information about a finite group which uses infinite groups in some of its steps. Because of this blurring of the dividing line between algorithms for finite groups and algorithms for infinite groups, it seems natural to discuss most of the known group-theoretic algorithms within the context of finitely presented groups.

In my first talk [15] I pointed out that many of the obvious questions which can be asked about a finitely presented group do not have algorithmic solutions. In the first part of this lecture I want to describe a few positive results which show that one can obtain useful information about an arbitrary finitely presented group. The second part of the talk will be devoted to a sketch of an algorithm for determining the order of the group generated by a given set of permutations of a finite set. This algorithm involves some computation with infinite groups. It is quite efficient even when the degree of the permutations is in the thousands.

Let us first fix some notation. Suppose G is a group containing the elements g and h. The *commutator* (g, h) is the element $g^{-1}h^{-1}gh$. If H and K are subgroups of G, then the *commutator subgroup* (H, K) is the subgroup of G

generated by all (h, k) with h in H and k in K. The *derived group* of G is the subgroup $G' = (G, G)$ and G'' is defined to be $(G')'$. Also, for any positive integer m the subgroup of G generated by all mth powers g^m of elements g in G will be denoted G^m.

Free groups should be familiar to most members of the audience. However, in this lecture I shall be discussing free monoids (semigroups with identity) and the connection between them and free groups. It therefore seems best to review the basic definitions.

Let X be a set. The *free monoid* generated by X is the set S of finite sequences, including the empty sequence, of elements from X. Elements of S are called *words*. The product of two words U and V is the sequence U, V obtained by following U by V. Given any map f of X into a monoid T, there is a unique extension of f to a homomorphism of S into T.

To construct the free group generated by X we first form the cartesian product $X^{\pm} = X \times \{1, -1\}$ and define S to be the free monoid generated by $X^{\pm}$. Following accepted practice, we shall denote the element (x, ε) of $X^{\pm}$ by x^{ε} and frequently identify x^1 with x. We shall also identify x^{ε} with the word of length 1 whose single term is x^{ε}. Thus we have $X \subseteq X^{\pm} \subseteq S$. We define $\sim$ to be the finest equivalence relation on S such that $U, x^{\varepsilon}, x^{-\varepsilon}, V$ is equivalent to U, V for all U and V in S and all x^{ε} in $X^{\pm}$. Let F denote the set of equivalence classes of $\sim$. For U in S we shall denote by $[U]$ the element of F containing U. The formula $[U][V] = [U, V]$ gives a well defined binary operation on F and with respect to this operation F is a group, the *free group* generated by X. There is an antiautomorphism $'$ of S defined as follows: For $u = x^{\varepsilon}$ in $X^{\pm}$ set $u' = x^{-\varepsilon}$ and for $U = u_1, \ldots, u_r$ in S set $U' = (u_r)', \ldots, (u_1)'$. The inverse of $[U]$ in F is $[U']$.

The homomorphisms of F into a group G are in 1-1 correspondence with the maps of X into G. They are also in 1-1 correspondence with the homomorphisms of S into G for which x^{ε} and $x^{-\varepsilon}$ are mapped to elements which are inverses of each other.

If R is a subset of S, then $N(R)$ will denote the subgroup of F generated by all conjugates of the elements $[R]$ with R in R. Thus $N(R)$ is the smallest normal subgroup of F containing all the $[R]$. The quotient group $F/N(R)$ is denoted $\langle X|R\rangle$. The pair X, R is called a presentation for any group isomorphic to $\langle X|R\rangle$. The presentation is finite if both X and R are finite. It is important to note that even when X and R are finite the subgroup $N(R)$ is not finitely generated unless $\langle X|\mathcal{R}\rangle$ is a finite group.

One of the reasons for stressing the role of S in the above definitions is the fact that algorithms for working with finitely presented groups actually manipulate words, that is, elements of S.

Let us now fix a finite set X with d elements and a finite subset $\mathcal{R}$ of the free monoid S generated by $X^{\pm}$. For concreteness we may take $X = \{1, \ldots, d\}$. We shall be discussing algorithms for determining properties of the group $G = \langle X|\mathcal{R}\rangle$. There are natural homomorphisms from S to F and F to $G = F/N(\mathcal{R})$. For U in S we shall denote by $\overline{U}$ the image of U under the composition of these maps. Thus $\overline{U}$ is the coset $N(\mathcal{R})[U]$. Although we can not hope to be able to describe the complete structure of G, we can obtain some useful information.

For example, it is possible to describe the abelian group G/G'. Let $e_1, \ldots, e_d$ be the standard basis of $\mathbb{Z}^d$. The map $i \mapsto e_i$ of X into $\mathbb{Z}^d$ defines a homomorphism f of F onto $\mathbb{Z}^d$. It is not hard to show that G/G' is isomorphic to $\mathbb{Z}^d/M$, where M is the image of $N = N(\mathcal{R})$ under f. Although N need not be finitely generated, it is easy to see that M is generated by the images under f of the elements $[R]$ with R in $\mathcal{R}$. Thus we have a finite generating set for M and we can use the reduction to Smith normal form discussed in [15] to find the orders of the cyclic factors of $\mathbb{Z}^d/M$.

It might seem a reasonable next step to try to describe the structure of G/G''. However, G/G'' need not be finitely presented. This situation occurs, for example, when $\mathcal{R}$ is empty. In this case $G = F$ and it is known that F/F'' is not finitely presented for $d > 1$.

Actually, the appropriate generalizations of the quotient group G/G' are the groups $G/\gamma_i(G)$, where $\gamma_i(G)$ is the ith term in the lower central series of G. Here $\gamma_0(G)$ and $\gamma_1(G)$ are defined to be G and for $i \geq 1$ we take $\gamma_{i+1}(G)$ to be the commutator subgroup $\left(G, \gamma_i(G)\right)$. (These subgroups were also discussed by Professor Wall [16].) We say that G is nilpotent of class c, c a nonnegative integer, if $\gamma_{c+1}(G) = 1$ and either $c = 0$ or $\gamma_c(G) \neq 1$. The group $G/\gamma_{c+1}(G)$ is nilpotent of class at most c.

If H is a finitely generated nilpotent group, then every subgroup of H is finitely generated. Moreover, if we have a finite presentation for a group H which is known to be nilpotent, then we can solve the word problem and the conjugacy problem for H. (For the conjugacy problem see [3].) For our group

$G = \langle X|R \rangle$, the quotient $G/\gamma_{c+1}(G)$ is a finitely presented nilpotent group and we can determine a great deal of its structure. It should be noted, however, that the isomorphism problem for finitely presented nilpotent groups has not been solved. Thus we do not yet know how to decide whether $G/\gamma_{c+1}(G)$ is isomorphic to a given nilpotent group H .

Although we know in principle how to compute such things as the orders of the cyclic direct factors of the abelian groups $\gamma_i(G)/\gamma_{i+1}(G)$, no general purpose computer program for doing this yet exists. Algorithms have been implemented which are designed to compute nilpotent quotient groups of G which are p-groups for some prime p . The various procedures for accomplishing this are lumped together under the term *nilpotent quotient algorithm*. For these purposes it seems useful to replace the lower central series by the lower exponent-p-central series, the series of groups $\gamma_i^p(G)$ defined as follows: set $\gamma_1^p(G) = G$ and if $H = \gamma_i^p(G)$, $i \geq 1$, then set $\gamma_{i+1}^p(G) = (G, H)H^p$. The groups $G/\gamma_i^p(G)$ are finite p-groups. Quite powerful programs for computing the orders of these quotients are available. A more complete discussion of the nilpotent quotient algorithm can be found in [10].

One application of the nilpotent quotient algorithm has been to various special cases of the Burnside problem. For a positive integer k let $B(d, k)$ be the group F/F^k , the largest d-generator group of exponent k . In [1] it is stated that $B(4, 4)$ has order 2^{422} and in [8] it is proved that the largest finite quotient group of $B(2, 5)$ has order 5^{34} . The proofs of both of these results involve computation with the nilpotent quotient algorithm. Since the groups $B(d, k)$ are not finitely presented as defined, a word of explanation is in order. As an example, let us look at the result about $B(2, 5)$ a little more carefully. It is possible to exhibit a finitely presented two-generator group B with the following properties:

1. B has $B(2, 5)$ as a homomorphic image;
2. $\gamma_{13}^5(B) = \gamma_{14}^5(B)$;
3. $B/\gamma_{13}^5(B)$ has exponent 5 and order 5^{34} .

From this it is easy to see that $B/\gamma_{13}^5(B)$ is the largest finite homomorphic image of $B(2, 5)$.

Another application of the nilpotent quotient algorithm has been the construction of all p-groups of a particular order satisfying some given property.

For example, in [2] all two-generator groups of order 3^8 and nilpotency class 6 are constructed. A general discussion of the way the nilpotent quotient algorithm can be used to construct p-groups can be found in [9].

The next few algorithms I wish to discuss are all related to subgroups of finite index in finitely presented groups. Two fundamental results in this area are given in the following theorems:

THEOREM 1. *Let H be a finitely presented group and let K be a subgroup of finite index in H. Then K has a finite presentation.*

THEOREM 2. *Let H be a finitely generated group. For each positive integer m there are only finitely many subgroups K of H with $|H : K| = m$.*

Theorem 1 is due to Reidemeister [11] with improvements in the proof by Schreier [12]. Theorem 2 is more elementary and has been known for a long time. The proofs of both theorems are constructive. The algorithms derived from the proofs of Theorem 1 and Theorem 2 are called the Reidemeister-Schreier and low index subgroup algorithms, respectively. In this talk there is only time enough to outline the main ideas of these algorithms. A description of one implementation of the Reidemeister-Schreier algorithm can be found in [7] while [5] describes an implementation of the low index subgroup algorithm.

Let H be a subgroup of index m in $G = \langle X|R \rangle$. Choose right coset representatives $u_1, \ldots, u_m$ for H in G. We shall assume that $u_1 = 1$ is the representative for H. For g in G let $\sigma(g)$ be the map of $\Omega = \{1, \ldots, m\}$ into itself taking i in Ω to j where $Hu_i g = Hu_j$. Then $\sigma(g)$ is actually an element of the symmetric group Σ_m and $\sigma : G \to \Sigma_m$ is a homomorphism. Given σ we can reconstruct H as the set of all g in G such that $\sigma(g)$ fixes 1. Moreover, if σ is any homomorphism of G into Σ_m such that $\sigma(G)$ is transitive, then $H(\sigma) = \{g \in G \mid 1^{\sigma(g)} = 1\}$ is a subgroup of G of index m. We have seen that every subgroup of index m in G occurs as $H(\sigma)$ for some σ. Unfortunately, it is entirely possible for $H(\sigma)$ to equal $H(\tau)$ for different homomorphisms σ and τ. Thus we can list the subgroups of index m in G provided we can solve the following problems:

1. List the set $\mathcal{H}$ of homomorphisms σ of G into Σ_m such that $\sigma(G)$ is transitive.

2. For σ and τ in $\mathcal{H}$ describe a procedure for deciding whether $H(\sigma) = H(\tau)$.

A homomorphism $\sigma : G \to \Sigma_m$ is determined by the images of the generators $\bar{x}$ with x in X. Given a map $s : X \to \Sigma_m$ we can extend s first to $X^{\pm}$ by defining $s(x^{-1})$ to be $s(x)^{-1}$ and then extend s to a homomorphism of S into Σ_m. We get a well defined homomorphism σ of G into Σ_m by setting $\sigma(\overline{U}) = s(U)$ if and only if $s(R) = 1$ for all R in R. Thus the homomorphisms of G into Σ_m are in 1-1 correspondence with the maps of X into Σ_m such that the images of the elements of x satisfy the defining relators in R. Since X and Σ_m are finite, there are only finitely many maps $s : X \to \Sigma_m$, and since R is finite we can decide which of these satisfy $s(R) = 1$ for all R in R. Thus we can list the homomorphisms σ of G into Σ_m. Deciding whether $\sigma(G)$ is transitive is also easy. Therefore we can solve Problem 1. For each σ in H we shall assume we have $\sigma(\overline{x})$ explicitly given for each x in X. Hence given a word U we can compute $\sigma(\overline{U})$. In particular, we can see whether $\sigma(\overline{U})$ fixes 1 and so decide whether $\overline{U}$ is in $H(\sigma)$.

One way to solve Problem 2 is to produce a finite generating set for $H(\sigma)$. Let us assume σ is fixed. For each i in Ω choose a word U_i in S such that $\sigma(\overline{U}_i)$ maps 1 to i. Since $\sigma(G)$ is transitive and is generated by the $\sigma(\overline{x})$, we can do this effectively. We assume that U_1 is the empty word. For i in Ω and x in X let $h(i, x) = U_i, x, U_j'$, where $j = i^{\sigma(\overline{x})}$. It is not hard to show (see Lemma 7.22 of [6]) that $H(\sigma)$ is generated by the $h(i, x)$, which are referred to as Schreier generators for $H(\sigma)$. Now we can solve Problem 2. If τ is another homomorphism in H, then to decide whether $H(\sigma) = H(\tau)$ we need only decide whether each $h(i, x)$ is in $H(\tau)$. But we have already remarked that τ determines $H(\tau)$ effectively. Thus we can find a subset H_0 of H such that every subgroup of index m in G is $H(\sigma)$ for exactly one σ in H_0.

The low index subgroup algorithm in the form just outlined is too inefficient to be of practical use. However, a more careful analysis of the computations involved has led to computer programs which can be used on interesting problems. It turns out that in order to find all subgroups of index m the programs go through essentially all the work of finding all subgroups of index not exceeding m. Thus the programs normally accept as input the presentation X, R and an integer n and produce a list of all subgroups of $\langle X|\mathrm{R}\rangle$ having index at most n. The values of n which are feasible depend heavily on the presentation. In some cases $n = 50$ can be handled easily and in other cases $n = 10$ proves very difficult.

The Reidemeister-Schreier algorithm allows us to compute a presentation for $H(\sigma)$ in terms of the Schreier generators $h(i, x)$. More precisely, let $Y = \Omega \times X$ and let T be the free monoid generated by $Y^{\pm}$. The map taking $(i, x)^{\varepsilon}$ in $Y^{\pm}$ to $h(i, x)^{\varepsilon}$ in $H(\sigma)$ extends to a homomorphism of T onto $H(\sigma)$. For A in T

let $\hat{A}$ denote the image of A in $H(\sigma)$. The Reidemeister-Schreier algorithm constructs a finite subset S of T such that Y, S is a presentation for $H(\sigma)$.

Every element g of G can be written uniquely in the form $h\overline{U}_i$ for some h in $H(\sigma)$ and some i in Ω. Thus given a word U in S and an integer i in Ω, there is a word A in T and an integer j such that $\overline{U}_i\overline{U} = \hat{A}\overline{U}_j$. The integer j is uniquely determined but the word A is not. The following lemma shows how to choose A in a "uniform" manner.

LEMMA. *Let* $U = \Omega \times S$ *and* $V = T \times \Omega$. *There is a unique map* $f : U \to V$ *such that*

(a) *if* $i \in \Omega$, $x \in X$, $j = i^{\sigma(\overline{x})}$ *and* $k = i^{\sigma(\overline{x})^{-1}}$, *then* f *maps* (i, x) *to* $\big((i, x), j\big)$ *and* $\big(i, x^{-1}\big)$ *to* $\big((k, x)^{-1}, k\big)$;

(b) *if* U *and* V *are in* S, i *is in* Ω *and* $W = U, V$, *then* f *maps* (i, W) *to* (C, k), *where* $C = A, B$, $f(i, U) = (A, j)$ *and* $f(j, V) = (B, k)$.

If $f(i, U) = (A, j)$, *then* $\overline{U}_i\overline{U} = \hat{A}\overline{U}_j$.

The proof of the lemma is straight forward.

For each R in R we have $\overline{R} = 1$ and so for i in Ω we know $\overline{U}_i\overline{R} = \overline{U}_i$. Thus $f(i, R) = (A, i)$ with $\hat{A} = 1$. Let S_1 be the set of relators A for $H(\sigma)$ obtained in this way. Now $h(i, x)$ is defined to be $\overline{U_i, x, U'_j}$, where $j = i^{\sigma(\overline{x})}$. Since $\overline{U}_1 = 1$, we have $f\big(1, (U_i, x, U'_j)\big) = (B, 1)$, where $\hat{B} = h(i, x)$. Thus $B, (i, x)^{-1}$ is a relator for $H(\sigma)$. Let S_2 be the set of these relators and set $S = S_1 \cup S_2$. Then the pair Y, S is a presentation for $H(\sigma)$.

If $|G : H(\sigma)| = m$, then Y has md elements and S has $m(r+d)$ elements, where $r = |R|$. If the set of words U_i is chosen to be a *Schreier system*, which means that whenever U, V is in the set then U is in the set too, then we can eliminate $m - 1$ of the generators in Y and the relators in S_2. This gives us a presentation for $H(\sigma)$ with $1 + m(d-1)$ generators and mr relators. Even with this improvement, the presentations obtained when m is fairly large, say $m \geq 100$, require considerable further processing to be useful. Some techniques for simplifying the presentations exist but they are mostly *ad hoc*.

One fairly common application of the Reidemeister-Schreier algorithm is to compute a presentation for a subgroup H and then use the first algorithm discussed above to compute the orders of the cyclic direct factors of H/H'. If H/H' happens to have a cyclic factor of infinite order, then H is infinite and thus so is G.

In this way one can prove that certain finitely presented groups are infinite.

It may appear from the discussion so far that we have pretty good control over the subgroups of finite index in a finitely presented group. However, things are not as nice as they look. Suppose we are given words $V_1, \ldots, V_t$ in S. It is natural to ask whether the subgroup H of G generated by $\overline{V}_1, \ldots, \overline{V}_t$ has finite index. Unfortunately, this question has no algorithmic solution. The exact situation is the following: there is no algorithm which, when H has infinite index, will verify this fact in a finite number of steps. There is an algorithm which, when $|G : H|$ is finite, will terminate and give the value of $|G : H|$. However, there is no way of giving an *a priori* estimate for the time needed to compute $|G : H|$ when it is finite.

Although we can not in general decide whether a given finitely generated subgroup of G has finite index or not, it is estimated that more computer time is spent on problems of this type than on any other computational problem in group theory. I want to describe next the algorithm used to try to prove the finiteness of $|G : H|$ for $H = \langle \overline{V}_1, \ldots, \overline{V}_t \rangle$.

Let H be any subgroup of $G = \langle X | R \rangle$. The set of right cosets of H in G is at most countably infinite. If H has infinite index, then let Ω be the set of positive integers. If $|G : H| = m < \infty$, then set $\Omega = \{1, \ldots, m\}$. For each i in Ω choose an element u_i of G such that $\{u_i \mid i \in \Omega\}$ is a set of right coset representatives for H in G. We shall assume $u_1 = 1$. As before we have a homomorphism σ of G into the symmetric group Σ on Ω given by $i^{\sigma(g)} = j$ if $Hu_i g = Hu_j$. Let us imagine a table whose columns are indexed by $X^{\pm}$ and whose rows are indexed by Ω such that the entry in row i and column u is $i^{\sigma(\overline{u})}$. For example, if $X = \{x, y\}$, then the table might look like

	x	x^{-1}	y	y^{-1}
1	3	2	6	8
2	1	4	7	5
3	4	1	9	9
4	2	3	8	6
5	8	6	2	11
6	5	7	4	1
7	6	8	10	2
8	7	5	1	4
9	10	11	3	3
10	13	9	12	7
⋮	⋮	⋮	⋮	⋮

The columns headed x^{-1} and y^{-1} are determined by the columns headed x and y, respectively, but for reasons of computational efficiency it is considered a good idea to have all four columns. If $|\Omega|$ is infinite, or even a very large finite number, then we can not possibly write down the whole table. We can, however, write down portions of the table. Let n be a moderately large positive integer, say $n \leq 100{,}000$. We can *truncate* the table after n cosets by taking only the first n rows and replacing any entries in these rows which are larger than n by a zero. For example, truncating the table above after six cosets gives the following:

	x	x^{-1}	y	y^{-1}
1	3	2	6	0
2	1	4	0	5
3	4	1	0	0
4	2	3	0	6
5	0	6	2	0
6	5	0	4	1

Each column of the truncated table defines a map of $\{1, \ldots, n\}$ into $\Delta_n = \{0, 1, \ldots, n\}$. If we agree that 0 is always to be mapped to 0, then each column defines a map of Δ_n into itself. The set M_n of all maps of Δ_n to itself which fix 0 is a monoid whose group of invertible elements is isomorphic to Σ_n. The truncated table thus defines a map f of $X^{\pm}$ into M_n. We can extend f uniquely to a homomorphism of S into M_n. Note that we do not have in general an associated homomorphism of F into M_n since it is possible to have words U and V with $U \sim V$ and $f(U) \neq f(V)$. In the example above we may take $U = x^{-1}$ and $V = x^{-1}, y, y^{-1}$. Then $f(U)$ maps 1 to 2 while $f(V)$ maps 1 to 0.

The homomorphisms $f : S \to M_n$ obtained by truncating the action of the generators and their inverses on the set of right cosets of a subgroup satisfy the following important condition:

I. Suppose U is in S and i is in Δ_n. If $f(U)$ maps i to j and $j \neq 0$, then $f(U')$ maps j to i.

If the coset representatives u_i are chosen in a "reasonable" way, then the following condition can also be made to hold:

II. For any i in Δ_n with $i \neq 0$ there exists a word U in S such that $f(U)$ maps 1 to i.

To insure that II holds we insist that elements u_i with i small be expressible as $\overline{U}_i$ where U_i is a short word in S.

A *coset table representation* of S is a homomorphism $f : S \to M_n$ satisfying conditions I and II. The term coset table refers to the matrix giving the values $i^{f(u)}$ for $1 \le i \le n$ and u in $X^{\pm}$. To check condition I it is sufficient to consider only those words U of length 1.

Suppose we are given a coset table representation $f : S \to M_n$ and a finitely presented group $G = \langle X|R \rangle$. Is it possible to decide whether f comes from truncating the action of G on the cosets of some subgroup? That is, can we find a subgroup H of G and a set of elements $1 = u_1, u_2, \ldots, u_n$ of G lying in distinct right cosets of H such that for U in S and $1 \le i \le n$ we have $j = i^{f(U)} \ne 0$ if and only if $Hu_i\overline{U} = Hu_j$? For some presentations we can give necessary and sufficient conditions but in general we have only the following necessary condition:

III. Suppose for some words U and V and some x^{ε} in $X^{\pm}$ the word U, x^{ε}, V is in R. If for some i in Δ_n both $j = i^{f(U)}$ and $k = i^{f(V')}$ are nonzero, then $f(x^{\varepsilon})$ maps j to k.

Condition III comes from the fact that for each R in R we must have $Hu_i\overline{R} = Hu_i$.

If $f : S \to M_n$ is a coset table representation satisfying III, then we can ask a further question. Given a finite subset W of S, could f be obtained by truncating the action of G on the cosets of a subgroup H which contains $\overline{W}$ for all W in W? Again all we have is the following necessary condition which is based on the fact that each $\overline{W}$ must fix the first coset of any such H:

IV. Suppose for some words U and V and some x^{ε} in $X^{\pm}$ the word U, x^{ε}, V is in W. If $j = 1^{f(U)}$ and $k = 1^{f(V')}$ are both nonzero, then $f(x^{\varepsilon})$ maps j to k.

The *Todd-Coxeter* or *coset enumeration* algorithm is an algorithm TC which takes as input X, two subsets R and W of S and a positive integer N. Suppose $G = \langle X|R \rangle$ and H is the subgroup of G generated by the set of $\overline{W}$ with W in W. Then the output of TC is a coset table representation $f : S \to M_n$ such that

A. Conditions III and IV are satisfied

B. If U and V are in S and $0 \ne 1^{f(U)} = 1^{f(V)}$, then $H\overline{U} = H\overline{V}$;

C. The integer n does not exceed N. If $n < N$, then $f(S)$ is contained in the group of invertible elements of M_n. (It follows that $n = |G : H|$ in this case.)

D. Let $d = |X|$ and let l be the sum of the lengths of the words in $R \cup W$. There exists a real valued function $t(d, l, N)$, such that if U is in S and the length of U does not exceed $t(d, l, N)$, then $1^{f(U)} \neq 0$. Moreover for fixed d and l, $t(d, l, N)$ tends monotonically to ∞.

Conditions A-D do not uniquely determine TC and so coset enumeration should be thought of as a family of algorithms. The parameter N should be thought of as limiting the space available to the algorithm. Todd-Coxeter algorithms terminate in a predictable time but the best upper bounds we know are exponential in N. In practice a much faster termination is experienced and more work on the speed of these algorithms needs to be done.

The following theorem formalizes the statement made earlier that when a finitely generated subgroup H of G has finite index we can compute $|G : H|$ but not in a predictable time.

THEOREM 3. *Let X, R and W be given and assume $H = \langle \overline{W} \mid W \in W \rangle$ has finite index in $G = \langle X|R \rangle$. There exists an integer N_0 such that for all $N \geq N_0$ the Todd-Coxeter algorithm terminates, when applied to the inputs X, R, W and N, with $n = |G : H|$.*

Although Theorem 3 states that given enough space the Todd-Coxeter algorithm will determine $|G : H|$, there is no way of effectively bounding N_0 in terms of some reasonable measure of the size of the input data X, R and W.

The computer implementation of the Todd-Coxeter algorithm has a long history. The best survey of the various approaches which have been tried can be found in [4].

Although we have seen several examples of group theoretic algorithms, we have not yet had any concrete examples of computational problems in group theory. It is now time to remedy this situation. Let x and y be the following elements of Σ_{15}:

$$x = (1, 15, 8)(2, 9, 10)(3, 11, 5)(4, 7, 14)(6, 13, 12),$$

$$y = (1, 10, 4, 13, 3)(2, 7, 12, 11, 15)(5, 8, 9, 14, 6).$$

There is clearly a finite procedure for determining the order of the group G generated by x and y. I doubt that many of you can see immediately what the answer is. However, if I tell you that $x^3 = y^5 = (x, y) = 1$, where (x, y) is the commutator $x^{-1}y^{-1}xy$, you can verify these facts in your head and then deduce that, since G is commutative, the order of G must be 15.

The preceding example was given in order to illustrate the following point: Given a set X of permutations of the finite set Ω, a good way to determine the order of the group generated by X is to look for short relations satisfied by the elements of X . I shall now sketch an algorithm based on this idea.

Let us consider another example, this time in Σ_9 . Let

$$x = (1)(2, 4, 5)(3, 6, 7)(8)(9) ,$$

$$y = (1, 2, 3)(4, 7, 8)(5, 6, 9) ,$$

and $G = \langle x, y \rangle$. It is easy to see that G is transitive and so the index of the stabilizer G_1 of 1 in G is 9 . Let $X = \{x, y\}$ and let S be the free monoid generated by $X^{\pm}$. Identifying the coset $G_1 g$ with the point 1^g , we get a coset table representation $f : S \to M_9$ defined by the following table:

	x	x^{-1}	y	y^{-1}
1	1	1	2	3
2	4	5	3	1
3	6	7	1	2
4	5	2	7	8
5	2	4	6	9
6	7	3	9	5
7	3	6	8	4
8	8	8	4	7
9	9	9	5	6

The permutation x fixes 1 and so $G_1 \supseteq \langle x \rangle$. Clearly $\langle x \rangle$ has order 3 and thus $|G| = 9|G_1| \geq 27$. Moreover, we have equality if and only if $|G : \langle x \rangle| = 9$.

For the subgroup G_1 we know the index but not the order and for $\langle x \rangle$ we know the order but not the index. If we had a subset R of S such that X, R was a presentation for G , then we could use coset enumeration to determine $|G : \langle x \rangle|$. However, the only relation we know so far is $x^3 = 1$, which gives a presentation for $\langle x \rangle$. We start by setting $R = \{x^3\}$ and try to add relators to R in order to get a presentation for G .

To find a new relator, we apply our favorite version of the Todd-Coxeter algorithm to the input $X, R, W = \{x\}$ and $N = 10$. The output is not uniquely determined but let us say it is the representation $f_1 : S \to M_{10}$ given by the following matrix:

	x	x^{-1}	y	y^{-1}
1	1	1	2	3
2	4	5	6	1
3	7	8	1	9
4	5	2	10	0
5	2	4	0	0
6	0	0	0	2
7	8	3	0	0
8	3	7	0	0
9	0	0	3	0
10	0	0	0	4

At this point the group $H = \langle X|R\rangle$ is infinite and the subgroup of H generated by x has infinite index but it turns out that comparing f and f_1 leads to useful information. We choose words $U_1, \ldots, U_{10}$ in S such that $f_1(U_i)$ maps 1 to i. One such choice would be

$$U_1 = \emptyset , \qquad U_6 = y^2 ,$$

$$U_2 = y , \qquad U_7 = y^{-1}x ,$$

$$U_3 = y^{-1} , \qquad U_8 = y^{-1}x^{-1} ,$$

$$U_4 = yx , \qquad U_9 = y^{-2} ,$$

$$U_5 = yx^{-1} , \qquad U_{10} = yxy .$$

The ten values $1^{f(U_i)}$ all lie in the set $\{1, \ldots, 9\}$ and so there exist U_i and U_j with $i \neq j$ such that $f(U_i)$ and $f(U_j)$ map 1 to the same point. By direct computation we find that $f(U_3)$ and $f(U_6)$ both map 1 to 3. This simply says that in G we have $1^{y^{-1}} = 1^{y^2}$, or equivalently, $1^{y^3} = 1$. We now compute y^3 in G. If $G_1 = \langle x\rangle$, then y^3 must be in $\langle x\rangle$. Actually we find that $y^3 = 1$ and so we have produced a new relator.

We now add the word y^3 to R and repeat the process, applying the Todd-Coxeter algorithm to X, R, W, N. Say the output is $f_2 : S \to M_{10}$ given by

	x	x^{-1}	y	y^{-1}
1	1	1	2	3
2	4	5	3	1
3	6	7	1	2
4	5	2	8	9
5	2	4	10	0
6	7	3	0	0
7	3	6	0	0
8	0	0	9	4
9	0	0	4	8
10	0	0	0	5

We again look for words U and V such that $f_2(U)$ and $f_2(V)$ map 1 to different nonzero points but $f(U)$ and $f(V)$ agree on 1. One such pair is $U = y^{-1}x^{-1}$ and $V = yxy$. This means that $yxyxy$ is in G_1 but is probably not in the subgroup generated by x in the group $\langle X|R\rangle$. Evaluating $yxyxy$ in G, we find $yxyxy = x^{-1}$ or $(xy)^3 = 1$.

Adding $(xy)^3$ to R and doing another coset enumeration gives us $f_3 : S \to M_{10}$ defined by

	x	x^{-1}	y	y^{-1}
1	1	1	2	3
2	4	5	3	1
3	6	7	1	2
4	5	2	7	8
5	2	4	9	10
6	7	3	0	0
7	3	6	8	4
8	0	0	4	7
9	0	0	10	5
10	0	0	5	9

Now we find that $yx^{-1}yx^{-1}y$ is in G_1 but is probably not in the subgroup generated by x in $\langle X|R\rangle$. Computing $yx^{-1}yx^{-1}y$ in G, we see that $yx^{-1}yx^{-1}y = x$ or $(x^{-1}y)^3 = 1$. Adding this relator to R and doing one more coset enumeration leads to the conclusion that $\langle x\rangle$ has index 9 in $H = \langle X|R\rangle$. Among the relators in R is the relator x^3 and so in H the order of x is at most 3 and hence $|H| \le 27$. But H has G as a homomorphic image and so $|H| \ge 27$. Therefore H and G are isomorphic groups of order 27 and X, R is a presentation for G.

A formal description of the algorithm just outlined was first given in [14], where it was called the Schreier-Todd-Coxeter algorithm since it combines ideas due to Schreier with coset enumeration. An earlier algorithm described in [13] uses only basic properties of Schreier generators and is not practical for permutation groups on several thousand points. A very efficient implementation of the Schreier-Todd-Coxeter algorithm has been written by J. Leon and he is preparing a description of his program with sample results for publication. As an example, the program was given half a dozen permutations in Σ_{1782} which generate the Suzuki sporadic simple group. The program correctly computed the order of this group to be 448,345,497,600 in 10 seconds on an IBM 370/168 computer.

Since [14] is not generally available, I shall close with a brief recursive description of the Schreier-Todd-Coxeter algorithm *STC*. Let X be a set of permutations of the finite set Ω and let $G = \langle X \rangle$. If U is a word in the free monoid S generated by $X^{\pm}$, then $\overline{U}$ will denote the element of G defined by U. Let R be a finite subset of S, possibly empty, such that $\overline{R} = 1$ for all R in R. Let $\mathrm{A} = \alpha_1, \ldots, \alpha_r$ be a sequence of points in Ω, also possibly empty. The algorithm *STC* takes as input X, R and A. The algorithm adds elements of G to X, additional relators to R and additional points to the end of A until the only element of G fixing each point in A is the identity and X, R is a presentation for G. In the terminology of [13], A is a base for G and X is a strong generating set for G relative to A. In addition the algorithm computes the order M of G and constructs an algorithm P for deciding membership in G. The input to P is a permutation z of Ω. The output of P is an integer p which is 0 or 1 and a word U in S. If $p = 1$, then $z = \overline{U}$. If $p = 0$ then z is not in G and $\overline{U}$ is an element of G agreeing with z on an initial segment of A of maximum length.

The steps of *STC* are as follows:

1. If X contains a nonidentity element, then go to Step 3.

2. Leave X and A unchanged. Set $\mathrm{R} = \mathrm{R} \cup X$ and $M = 1$. Define the algorithm P to return $p = 0$ or 1 according as $z \neq 1$ or $z = 1$ and to return U as the empty word. Stop.

3. If $r = 0$, then choose a point α_1 in Ω which is moved by some element of X.

4. Let Y be the set of elements in X which fix α_1 and let S be the set of words in R which involve only terms from $Y^{\pm}$. Let $B = \alpha_2, \ldots, \alpha_r$. Apply *STC* to Y, S, B. Let $L = |\langle Y \rangle|$ and let Q be the membership algorithm for $\langle Y \rangle$.

5. Set $X = X \cup Y$ and $R = R \cup S$. Let A be the sequence α_1, B obtained by following α_1 by the terms in B .

6. Determine the orbit $\Delta = \alpha_1^G$ and for each δ in Δ choose a word $U(\delta)$ in S of minimal length such that $\overline{U(\delta)}$ maps α_1 to δ . Set $m = |\Delta|$.

7. Choose an integer $N > m$, say $N = m + 1$, and apply TC to X, R, Y and N . Let the result be the coset table representation $f : S \to M_n$.

8. If $n = m$, then go to Step 14.

9. Choose words U and V in S of minimal lengths such that $f(U)$ and $f(V)$ map 1 to different nonzero integers but $\overline{U}$ and $\overline{V}$ map α_1 to the same point.

10. Let $z = \overline{U}\overline{V}^{-1}$ and apply algorithm Q to z . Let the outputs be q and E .

11. If $q = 0$, then go to Step 13.

12. Add the word U, V', E' to R and go to Step 7.

13. Let $w = z\overline{E}^{-1}$. Add w to X and the word U, V', E', w' to R . Then go to Step 4.

14. Set $M = mL$ and define the algorithm P by the following steps:

    (a) if $\delta = (\alpha_1)^z$ is in Δ , then go to Step (c);

    (b) set $p = 0$ and U equal to the empty word. Stop;

    (c) apply Q to $z\overline{U(\delta)}^{-1}$ and let the outputs be p and V . Set $U = V, U(\delta)$ and stop.

15. Stop.

References

[1] William A. Alford, George Havas and Michael F. Newman, "Groups of exponent four", *Notices Amer. Math. Soc.* 22 (1975), A.301.

[2] Judith A. Ascione, George Havas, and C.R. Leedham-Green, "A computer aided classification of certain groups of prime power order", *Bull. Austral. Math. Soc.* 17 (1977), 257-274. Corrigendum and Microfiche Supplement, *Bull. Austral. Math. Soc.* 17 (1977), 317-320.

[3] Norman Blackburn, "Conjugacy in nilpotent groups", *Proc. Amer. Math. Soc.* 16 (1965), 143-148.

[4] John J. Cannon, Lucien A. Dimino, George Havas and Jane M. Watson, "Implementation and analysis of the Todd-Coxeter algorithm", *Math. Comp.* 27 (1973), 463-490.

[5] Anke Dietze and Mary Schaps, "Determining subgroups of a given finite index in a finitely presented group", *Canad. J. Math.* 26 (1974), 769-782.

[6] Marshall Hall, Jr., *The Theory of Groups* (Macmillan, New York, 1959).

[7] George Havas, "A Reidemeister-Schreier program", *Proc. Second Internat. Conf. Theory of Groups*, Canberra, 1973, 347-356 (Lecture Notes in Mathematics, 372. Springer-Verlag, Berlin, Heidelberg, New York, 1974).

[8] George Havas, G.E. Wall, and J.W. Wamsley, "The two generator restricted Burnside group of exponent five", *Bull. Austral. Math. Soc.* 10 (1974), 459-470.

[9] M.F. Newman, "Determination of groups of prime-power order, *Group Theory* (Proc. Miniconf., Canberra 1975, 73-84.. Lecture Notes in Mathematics, 573. Springer-Verlag, Berlin, Heidelberg, New York, 1977).

[10] M.F. Newman, "Calculating presentations for certain kinds of quotient groups", *SYMSAC '76* (ACM Sympos. Symbolic and Algebraic Computation, Yorktown Heights, New York, 1976, 2-8. ACM, New York, 1976). See also: Abstract, *Sigsam Bull.* 10 (1976), no. 3, 5.

[11] Kurt Reidemeister, "Knoten und Gruppen", *Abh. Math. Sem. Hamburg Univ.* 5 (1927), 7-23 (1926).

[12] Otto Schreier, "Die Untergruppen der freien Gruppen", *Abh. Math. Sem. Hamburg Univ.* 5 (1927), 161-183.

[13] Charles C. Sims, "Computation with permutation groups", *Proc. Second Sympos. Symbolic and Algebraic Manipulation*, Los Angeles, California, 1971, 23-28 (ACM, New York, 1971).

[14] Charles C. Sims, "Some algorithms based on coset enumeration" (unpublished, Nŏ. 1, 1974).

[15] Charles C. Sims, "The role of algorithms in the teaching of algebra", these proceedings, 95-107.

[16] G.E. Wall, "Lie methods in group theory", these proceedings, 137-173.

Department of Mathematics,
Rutgers University,
New Brunswick,
New Jersey,
USA.

PROC. 18th SRI 20-02
CANBERRA 1978, 125-130. (20D99)

A METHOD FOR CONSTRUCTING A GROUP FROM A SUBGROUP

Charles C. Sims

I have been involved in the construction of four sporadic finite simple groups, [1], [6], [4] and [3]. The first of these was almost literally constructed over dessert on the evening of September 2, 1967. However, in each of the other cases evidence for the existence of the groups was obtained long before the groups were shown to exist. Moreover, the existence proofs required extensive machine computation. The general pattern of these three constructions was the same. The simple group G, if it existed, was known to have a large subgroup H isomorphic to a group whose existence had already been established. In addition, there was known to be an element z of G such that $H \cap H^z$ was also fairly large. The method of construction of G was to choose a set Ω with $|\Omega| = |G : H|$ and to write down permutations of Ω corresponding to the way z and the elements of H ought to permute the right cosets of H in G. Once this had been done, there still remained the nontrivial problem of showing that the permutations of Ω so defined actually generated a group with the right properties. The main point I want to make is that since G was simple, describing the action of an element g of G on the set of right cosets of H actually determined g uniquely.

There are times when one wants to construct a group G and one knows a large subgroup H of G but the intersection of the conjugates of H in G is nontrivial and so the action of G on the right cosets of H is not faithful. For example, Fischer and Thompson and also Griess have obtained evidence for the existence of a simple group M of order

$$2^{46}.3^{20}.5^{9}.7^{6}.11^{2}.13^{3}.17.19.23.29.31.41.47.59.71$$

or about 8.08×10^{53} . The group M is commonly referred to as the Monster. If M exists, then the largest proper subgroup of M is almost certainly a group H which is a two-fold covering group of Fischer's Baby Monster, whose existence was announced in [3]. Unfortunately there is as yet no proof of the existence of this covering group which is independent of the existence of M . Thus if we want to construct M from H , then we must first construct H .

The Baby Monster was constructed as a permutation group of degree 13,571,955,000 . The smallest degree of a faithful permutation representation of H is much larger than this and the construction of H along the lines of the earlier constructions would be very difficult. In an attempt to find a better way to construct H , I decided to look at the method by which the covering groups of the symmetric groups were shown to exist. These covering groups were first constructed by Schur [5], who wrote down complex matrices which generated the groups. A version of this construction is given in [2], where a covering group of Σ_n is described as the group K generated by a particular set of $n - 1$ matrices. To verify that K is in fact a covering group, one has to show that for each generator A we have $A^2 = I$ and for each pair of generators A and B either $(AB)^2 = -I$ or $(AB)^3 = -I$, where I is the identity matrix. Since the generators are given either as tensor products of 2-by-2 matrices or as a sum of two such tensor products, these calculations are not very hard to carry out.

It seems unlikely that one can write down generating matrices for the group H in a form which would permit verification of relations. The main purpose of this talk is to give an alternate construction of the covering group of Σ_n described in [2]. The methods involved appear to be applicable to the construction of H and should make it possible to prove the existence of H with roughly the same amount of work as was needed to construct the Baby Monster. This approach might possibly be used to construct M from H as well. The method to be described has the advantage that it is possible to estimate quite early how much computing is involved. With the method used to construct the Lyons and O'Nan groups and the Baby Monster a great deal more work has to be done before one can tell precisely what computing resources will be needed to complete the project.

Let G be any group. There is a homomorphism of $G \times G$ into the symmetric group $\Sigma(G)$ on G given by

$$g^{(h,k)} = h^{-1}gk$$

for all g, h and k in G . Let τ be the automorphism of $G \times G$ taking (h, k) to (k, h) and let t be the inverse map on G , that is, $g^t = g^{-1}$ for all g in

G . Then

$$g^{t(h,k)} = g^{(h,k)^\tau t}$$

and so t normalizes the image $\overline{G}$ of $G \times G$ in $\Sigma(G)$. The basic idea of the construction technique to be discussed is to use the methods of earlier constructions to build up the group $\langle \overline{G}, t \rangle$. It should be remarked that the map of $G \times G$ onto $\overline{G}$ need not be an isomorphism. The kernel is $\{(g, g) \mid g \in Z(G)\}$. However, restricted to $1 \times G$ the map is injective. Thus to construct G we shall not write down "formal right cosets" as was done before but rather "formal group elements".

Let n be a positive integer and let $\hat{\Sigma}_n$ be the group generated by the n elements $x_1, \ldots, x_{n-1}, d$ and defined by the relations

$$d^2 = x_i^2 = (dx_i)^2 = 1 , \quad 1 \le i \le n-1 ,$$

$$(x_i x_{i+1})^3 = d , \qquad 1 \le i < n-1 ,$$

$$(x_i x_j)^2 = d , \qquad 2 \le i+1 < j \le n-1 .$$

The element d is central in $\hat{\Sigma}_n$. Setting $d = 1$, we obtain the standard presentation for Σ_n in terms of the transpositions $(1, 2), (2, 3), \ldots, (n-1, n)$. Thus $\hat{\Sigma}_n/\langle d\rangle \cong \Sigma_n$. Since d has order at most 2 , we see that $|\hat{\Sigma}_n| \le 2n!$. The remainder of this talk will be devoted to giving a proof of the following theorem:

THEOREM. *The order of* $\hat{\Sigma}_n$ *is* $2n!$.

The proof will proceed by induction. It is trivial to verify that $\hat{\Sigma}_1 \cong \mathbb{Z}_2$, $\hat{\Sigma}_2 \cong \mathbb{Z}_2 \times \mathbb{Z}_2$ and $\hat{\Sigma}_3 \cong \Sigma_3 \times \mathbb{Z}_2$. Thus we shall assume $n \ge 3$ and $|\hat{\Sigma}_n| = 2n!$ and prove that $|\hat{\Sigma}_{n+1}| = 2(n+1)!$.

Let $H = \hat{\Sigma}_n$ and let K be the subgroup of H generated by $x_1, \ldots, x_{n-2}$ and d . Clearly K is a homomorphic image of $\hat{\Sigma}_{n-1}$. Our induction assumption implies that $K \cong \hat{\Sigma}_{n-1}$.

LEMMA 1. *There is a unique automorphism* ρ *of* K *such that* $d^\rho = d$ *and* $x_i^\rho = dx_i$, $1 \le i \le n-2$. *The order of* ρ *is* 2 .

Proof. The elements $dx_1, \ldots, dx_{n-2}, d$ generate K and satisfy the same defining relations as $x_1, \ldots, x_{n-2}, d$. Thus there is a unique ρ in $\operatorname{aut}(K)$ fixing d and mapping x_i to dx_i . Since

$$x_i^{\rho^2} = (dx_i)^{\rho} = ddx_i = x_i ,$$

we have $\rho^2 = 1$.

Now let H_1 be the diagonal subgroup of $H \times H$ and let H_2 be the subgroup of $H \times H$ consisting of the pairs (k, k^{ρ}) with k in K . For $i = 1, 2$ let Ω_i be the set of right cosets of H_i in $H \times H$ and let $\Omega = \Omega_1 \cup \Omega_2$. Then

$$|\Omega_1| = 2n! ,$$

$$|\Omega_2| = 2nn! ,$$

$$|\Omega| = 2(n+1)! .$$

The element $H_i(a, b)$ in Ω_i will be denoted $[a, b]_i$. Elements of $H \times H$ act on Ω by right multiplication. That is, we can define

$$[a, b]_i^{(u,v)} = [au, bv]_i .$$

The orbits of $H \times H$ on Ω are simply Ω_1 and Ω_2 . We shall eventually identify Ω with the set of elements of $\hat{\Sigma}_{n+1}$ in such a way that Ω_1 and Ω_2 are the double cosets of $\hat{\Sigma}_n$ in $\hat{\Sigma}_{n+1}$.

Now let τ be the automorphism of $H \times H$ such that $(a, b)^{\tau} = (b, a)$. Then τ clearly normalises H_1 . Moreover, if a is in K , then

$$(a, a^{\rho})^{\tau} = (a^{\rho}, a) = (a^{\rho}, (a^{\rho})^{\rho})$$

and so τ normalizes H_2 also. Thus τ maps cosets of H_i into cosets of H_i , $i = 1, 2$. In other words the formula

$$[a, b]_i^t = [b, a]_i$$

defines a permutation t of Ω with $t^2 = 1$. Under our identification of Ω and $\hat{\Sigma}_{n+1}$ the map t will turn out to be the inverse map.

At this point we would like to define a certain permutation z on Ω . To do so we shall invoke the following elementary but very important lemma:

LEMMA 2. *Let* X *be a permutation group on a set* Δ *and let* $\Delta_1, \ldots, \Delta_r$ *be the orbits of* X *. For* $1 \le i \le r$ *let* δ_i *be a representative of* Δ_i *. Suppose* σ *is an automorphism of* X *and* $\gamma_1, \ldots, \gamma_r$ *are points of* Δ *such that*

(a) if $i \neq j$, then γ_i and γ_j are in different X orbits;

(b) if $1 \leq i \leq r$, then

$$\left(X_{\delta_i}\right)^{\sigma} = X_{\gamma_i} .$$

Then there is a unique permutation z of Δ such that $z^{-1}xz = x^{\sigma}$ for all x in X and $\delta_i^z = \gamma_i$, $1 \leq i \leq r$.

Proof. Let us first show there is at most one such permutation z. If δ is in Δ, then $\delta = \delta_i^x$ for some i and some x in X. Thus

$$\delta^z = \delta_i^{xz} = \delta_i^{zz^{-1}xz} = \gamma_i^{x^{\sigma}} .$$

Therefore δ^z is uniquely determined. To show that z exists we must show that the formula

$$\left(\delta_i^x\right)^z = \gamma_i^{x^{\sigma}}$$

describes a well defined map of Δ into itself which is in fact a permutation. Suppose x and y are in X and $\delta_i^x = \delta_i^y$. Then xy^{-1} is in X_{δ_i} and so by assumption $\left(xy^{-1}\right)^{\sigma} = x^{\sigma}\left(y^{\sigma}\right)^{-1}$ in X_{γ_i}. Therefore

$$\gamma_i^{x^{\sigma}} = \gamma_i^{y^{\sigma}}$$

and so z is well defined. Clearly z maps Δ_i onto γ_i^X and since the γ_i are in distinct orbits of X, we see that z maps Δ onto Δ. Suppose x and y are in X and

$$\left(\delta_i^x\right)^z = \left(\delta_j^y\right)^z .$$

Then we must have $i = j$ and

$$\gamma_i^{x^{\sigma}} = \gamma_i^{y^{\sigma}} .$$

Thus $x^{\sigma}\left(y^{\sigma}\right)^{-1} = \left(xy^{-1}\right)^{\sigma}$ is in X_{γ_i} and so xy^{-1} is in X_{δ_i}. Therefore

$$\delta_i^x = \delta_i^y$$

and z is injective. Hence z is a permutation of Δ. Clearly $\delta_i^z = \gamma_i$. All that remains is to check the condition $z^{-1}xz = x^{\sigma}$, or equivalently, $xz = zx^{\sigma}$. Now if y is in X, then

$$\left(\delta_i^y\right)^{xz} = \left(\delta_i^{yx}\right)^z = \gamma_i^{(yx)^{\sigma}}$$

and

$$\left(\delta_i^y\right)^{zx^{\sigma}} = \gamma_i^{y^{\sigma}x^{\sigma}} .$$

Since σ is an isomorphism, the permutations xz and zx^{σ} agree on all points of Δ and so they are equal.

We shall now apply Lemma 2 to the situation in which $\Delta = \Omega$, X is the group induced by $H \times K$ and σ is the automorphism of X corresponding to the automorphism $1 \times \rho$ of $H \times K$. To start with we must make sure σ is really well defined. The kernel of our representation of $H \times H$ into $\Sigma(\Omega)$ is the cyclic group of order 2 generated by (d, d). This subgroup is contained in $H \times K$ and is fixed by $1 \times \rho$. Therefore $1 \times \rho$ induces an automorphism of $H{\times}K/\langle (d, d)\rangle$, which is isomorphic to X.

To apply Lemma 2 we need to know the orbits of $H \times K$ on Ω.

LEMMA 3. *The points* $\delta_1 = [1, 1]_1$, $\delta_2 = [1, 1]_2$ *and* $\delta_3 = [1, x_{n-1}]_2$ *are representatives for the orbits of* $H \times K$ *on* Ω.

Proof. Since

$$[a, b]_1 = [b, b]_1^{(b^{-1}a,1)} = \delta_1^{(b^{-1}a,1)} ,$$

$H \times K$ acts transitively on Ω_1. Also $H = K \cup Kx_{n-1}K$ and so any point in Ω_2 has the form $[h, k]_2$ or $[h, x_{n-1}k]_2$ with h in H and k in K. It is not hard to see that the points of each of these two types form an orbit of $H \times K$. Thus $H \times K$ has three orbits on Ω and δ_1, δ_2 and δ_3 are orbit representatives.

Let $\delta_4 = [dx_{n-1}, x_{n-1}]_2 = \delta_3^{(dx_{n-1},1)}$. A simple computation shows that

$(H \times K)_{\delta_1}$ = the diagonal of $K \times K$,

$(H \times K)_{\delta_2} = H_2$,

$(H \times K)_{\delta_3}$ = the diagonal of $L \times L$, where $L = \langle x_1, \ldots, x_{n-3}, d\rangle$,

$(H \times K)_{\delta_4} = \{(a, a^{\rho}) \mid a \in L\}$.

(Note that $x_{n-1}ax_{n-1} = a^\rho$ for all a in L .)

The automorphism $1 \times \rho$ of $H \times K$ interchanges the first and second and the third and fourth of these groups. Thus by Lemma 2 there is a unique permutation z of Ω such that $\delta_1^z = \delta_2$, $\delta_2^z = \delta_1$ and $\delta_3^z = \delta_4$ and

$$\delta^{(h,k)z} = \delta^{z(h,k^\rho)}$$

for all δ in Ω and all (h, k) in $H \times K$. In fact z is an involution. To see this we first observe that z^2 commutes with permutations of Ω induced by elements of $H \times K$. Thus to show $z^2 = 1$ it is enough to show z^2 fixes δ_1, δ_2 and δ_3 . Clearly z^2 fixes δ_1 and δ_2 . Now

$$\delta_3^{z^2} = \delta_4^z = \delta_3^{(dx_{n-1},1)z}$$
$$= \delta_3^{z(dx_{n-1},1)} = \delta_4^{(dx_{n-1},1)} = \delta_3 .$$

Thus $z^2 = 1$.

Let G be the subgroup of $\Sigma(\Omega)$ generated by the image of $1 \times H$ and z .

LEMMA 4. *The group G is transitive on Ω .*

Proof. Since

$$[a, b]_1 = [1, 1]_1^{(1,a^{-1}b)} ,$$

Ω_1 is a $1 \times H$ orbit. Now z maps Ω_1 onto

$$([1, 1]_2)^{H\times K} .$$

But

$$[a, b]_2 = ([a, 1]_2)^{(1,b)}$$

and $[a, 1]_2$ is in Ω_1^z while $(1, b)$ is in $1 \times H$. Thus G is transitive on Ω .

Since G is transitive, so is its conjugate $G^t = \langle H\times 1, z^t \rangle$. Our next goal is to show that G and G^t centralize each other. Once this is known, then Proposition 4.5 in Chapter I of [7] implies that G is regular and so $|G| = |\Omega| = 2(n+1)!$. It is then an easy matter to show that G is a homomorphic image of $\hat{\Sigma}_{n+1}$.

Clearly elements of $1 \times H$ and $H \times 1$ commute in their action on Ω . Also from the definition of z we know that z commutes elementwise with $H \times 1$ and so z^t commutes elementwise with $1 \times H = (H \times 1)^t$.

Thus to show G and G^t centralize each other it is enough to prove

LEMMA 5. *The permutations* z *and* z^t *commute. Equivalently,* $(tz)^4 = 1$.

Proof. Both t and z normalize the image of $K \times K$ in $\Sigma(\Omega)$. They induce the automorphisms corresponding to τ and $\sigma = 1 \times \rho$, respectively. If (a, b) is in $K \times K$, then

$$(a, b)^{\tau\sigma} = \left(b, a^{\rho}\right),$$

$$(a, b)^{(\tau\sigma)^2} = \left(a^{\rho}, b^{\rho}\right),$$

$$(a, b)^{(\tau\sigma)^4} = \left(a^{\rho^2}, b^{\rho^2}\right) = (a, b).$$

Thus $(tz)^4$ commutes with the image of $K \times K$. To show $(tz)^4 = 1$, it is enough to show that $(tz)^4$ fixes one point from each $K \times K$ orbit. Now $K \times K$ has seven orbits on Ω with the following representatives:

$$\beta_1 = [1, 1]_1, \qquad \beta_5 = \left[1, x_{n-1}\right]_2,$$

$$\beta_2 = \left[1, x_{n-1}\right]_1, \qquad \beta_6 = \left[dx_{n-1}, x_{n-1}\right]_2,$$

$$\beta_3 = [1, 1]_2, \qquad \beta_7 = \left[x_{n-1}, x_{n-2}x_{n-1}\right]_2.$$

$$\beta_4 = \left[x_{n-1}, 1\right]_2,$$

We shall omit the details of this computation. It is based on the fact that $H = K \cup \left(Kx_{n-1}K\right)$ and $K = L \cup \left(Lx_{n-2}L\right)$. To see that the number seven is right, we note that if we are going to be able to identify Ω with $\hat{\Sigma}_{n+1}$, then the orbits of $K \times K$ will be the (K, K)-double cosets in $\hat{\Sigma}_{n+1}$. Now d is in K and modulo d the group K corresponds to the stabilizer in Σ_{n+1} of the points $n + 1$ and n. Thus the number of double cosets of K is the number of orbits of Σ_{n+1} on ordered quadruples of points (i, j, k, l) with $i \neq j$ and $k \neq l$. There are seven such orbits with representatives

$$(n+1, n, n+1, n), \qquad (n+1, n, n-1, n+1),$$

$$(n+1, n, n+1, n-1), \qquad (n+1, n, n-1, n),$$

$$(n+1, n, n, n+1), \qquad (n+1, n, n-1, n-2).$$

$$(n+1, n, n, n-1),$$

It turns out that both t and z fix the set $\{\beta_1, \beta_2, \beta_3, \beta_4, \beta_5, \beta_6\}$. On this set t induces the permutation

$$(\beta_1)(\beta_2)(\beta_3)(\beta_4, \beta_5)(\beta_6) ,$$

while z induces

$$(\beta_1, \beta_3)(\beta_2, \beta_4)(\beta_5, \beta_6) .$$

Therefore, tz induces

$$(\beta_1, \beta_3)(\beta_2, \beta_4, \beta_6, \beta_5)$$

and so $(tz)^4$ fixes β_i , $1 \le i \le 6$. Thus all that remains is to show that $(tz)^4$ fixes β_7 . Let $\beta_8 = \beta_7^t = [x_{n-2}x_{n-1}, x_{n-1}]_2$. Then

$$\beta_8 = \beta_5^{(x_{n-2}x_{n-1},1)}$$

and so

$$\begin{aligned}
\beta_8^z = \beta_5^{z(x_{n-2}x_{n-1},1)} &= \beta_6^{(x_{n-2}x_{n-1},1)} \\
&= [dx_{n-1}x_{n-2}x_{n-1}, x_{n-1}]_2 \\
&= [x_{n-2}x_{n-1}x_{n-2}x_{n-1}, x_{n-2}x_{n-1}]_2 \\
&= [dx_{n-1}x_{n-2}, x_{n-2}x_{n-1}]_2 \\
&= \beta_7^{(x_{n-2},1)} .
\end{aligned}$$

Thus

$$\beta_7^{tz} = \beta_7^{(x_{n-2},1)} .$$

Since $(x_{n-2}, 1)$ is in $K \times K$,

$$\begin{aligned}
\beta_7^{(tz)^2} = \beta_7^{(x_{n-2},1)tz} &= \beta_7^{tz(1,dx_{n-2})} \\
&= \beta_7^{(x_{n-2},dx_{n-2})} .
\end{aligned}$$

Similarly,

$$\begin{aligned}
\beta_7^{(tz)^4} &= \beta_7^{(x_{n-2},dx_{n-2})(tz)^2} \\
&= \beta_7^{(tz)^2(dx_{n-2},x_{n-2})} \\
&= \beta_7^{(d,d)} = \beta_7 .
\end{aligned}$$

Therefore $(tz)^4 = 1$.

As noted above, we now know that G is regular. The elements $(1, x_i)$, $1 \le i \le n-1$, z and $(1, d)$ generate G . We must now show that these elements satisfy the defining relations for $\hat{\Sigma}_{n+1}$. Since $H \cong \hat{\Sigma}_n$, we need only check the relations

$$z^2 = [z(1, d)]^2 = 1 ,$$

$$[z(1, x_{n-1})]^3 = (1, d) ,$$

$$[z(1, x_i)]^2 = (1, d) , \quad 1 \le i \le n-2 .$$

We already know that $z^2 = 1$, that z commutes with $(1, d)$ and that $z(1, x_i)z = (1, dx_i)$ for $1 \le i \le n-2$. Thus we have only to show

$$[z(1, x_{n-1})]^3 = (1, d) .$$

Since G is regular, we need only verify that $[z(1, x_{n-1})]^3$ and $(1, d)$ agree on the point $\beta_1 = [1, 1]_1$. Now

$$\beta_1^z = \beta_3 ,$$

$$\beta_3^{(1,x_{n-1})} = \beta_5 ,$$

$$\beta_5^z = \beta_6 ,$$

$$\beta_6^{(1,x_{n-1})} = [dx_{n-1}, 1]_2 = \beta_4^{(1,d)} ,$$

$$\beta_4^{(1,d)z} = \beta_2^{(1,d)} ,$$

$$\beta_2^{(1,d)(1,x_{n-1})} = \beta_1^{(1,d)} .$$

Thus $[z(1, x_{n-1})]^3 = (1, d)$ and G is a homomorphic image of $\hat{\Sigma}_{n+1}$. Therefore, $|\hat{\Sigma}_{n+1}| \ge |G| = 2(n+1)!$, proving our theorem.

The same approach could be used to construct any group G from a known subgroup H provided there is an element z of G such that $G = \langle H, z\rangle$, z^2 is in H and $K = H \cap H^z$ does not have too many double cosets, since the difficult step, namely showing z and z^t commute, requires an amount of work proportional to the number of

(K, K)-double cosets in G. For example, suppose the existence of the two-fold covering group H of the Baby Monster has been established. There is in the Monster M an involution z such that $K = H \cap H^z$ has order approximately 3.06×10^{23}. A lower bound for the number of double cosets of K in M is

$$|M|/|K|^2 \sim 8.6 \times 10^6 .$$

It is not being unreasonable to hope that the number of double cosets of K in M is something like 10^8. A modern magnetic disk can store at least 10^8 bytes consisting of 8 binary digits. Thus it is quite possible that a description of a set of representatives for the (K, K)-double cosets could be stored on several such disks. The number of seconds in a year is a little more than 3×10^7. It is plausible to imagine that representatives for the orbits of $K \times K$ acting on the set of formal group elements can be found and processed at an average rate of at least 10 per second. These rough calculations indicate that the construction of the monster is technically feasible but not yet economically justifiable. I doubt that anyone would argue that the existence or nonexistence of the monster is of such importance to the mathematical community that it warrants devoting the resources of a major computer center for the better part of a year to settle the matter. However, it should be possible to refine the techniques sufficiently so that the question can be answered at a price which can be justified.

References

[1] Donald G. Higman and Charles C. Sims, "A simple group of order 44,352,000 ", *Math. Z.* **105** (1968), 110-113.

[2] B. Huppert, *Endliche Gruppen I* (Die Grundlehren der mathematischen Wissenschaften, **134**. Springer-Verlag, Berlin, Heidelberg, New York, 1967).

[3] Jeffrey S. Leon and Charles C. Sims, "The existence and uniqueness of a simple group generated by {3, 4}-transpositions", *Bull. Amer. Math. Soc.* **83** (1977), 1039-1040.

[4] Michael E. O'Nan, "Some evidence for the existence of a new simple group", *Proc. London Math. Soc.* **32** (1976), 421-479.

[5] J. Schur, "Über die Darstellung der symmetrischen und der alternierenden Gruppe durch gebrochene lineare Substitutionen", *J. reine angew. Math.* **139** (1911), 155-250.

[6] Charles C. Sims, "The existence and uniqueness of Lyons' groups", *Finite Groups '72* (Proc. Gainesville Conf., 1972, 138-141. North-Holland Mathematics Studies, **7**. North-Holland, Amsterdam, London; American Elsevier, New York, 1973).

[7] Helmut Wielandt, *Finite Permutation Groups* (translated from German by R. Bercov. Academic Press, New York, London, 1964).

Department of Mathematics,
Rutgers University,
New Brunswick,
New Jersey,
USA.

PROC. 18th SRI
CANBERRA 1978, 137-173.

20F40
(20F50, 17B60, 16A68)

LIE METHODS IN GROUP THEORY

G.E. Wall

This is an expanded version of some talks delivered at the Summer Research Institute. Their aim was to give an introduction to the use of Lie methods in group theory rather than a comprehensive survey of the subject. Thus, the choice of topics has been influenced by my own particular interests and important areas have been passed over without mention. Many of the proofs are either given in outline or omitted.

General references on the subject are Bourbaki [1, 2] and Magnus, Karrass, Solitar [12].

PART I

FILTRATIONS

1. Filtrations of rings and groups

Although the title refers to Lie rings and groups, associative rings[1] keep coming in too. Let me begin, then, with an associative algebra S over a commutative ring

[1] All associative rings (and algebras) under consideration are assumed to have identity elements. A subring of a ring R contains the 1 of R ; a homomorphism from R to a ring S maps the 1 of R to the 1 of S ; if M is an R-module, the 1 of R acts as the identity on M .

K , and a *filtration*[2] of S , that is, a descending chain of ideals

$$(1.1) \qquad \mathcal{S} : S = S^{(0)} \supseteq S^{(1)} \supseteq \ldots$$

such that

$$(1.2) \qquad S^{(r)}S^{(s)} \subseteq S^{(r+s)} \quad (r, s \geq 0) .$$

A typical example is $S \supseteq J \supseteq J^2 \supseteq \ldots$, where J is any ideal of S , and it is easy to make up others. An example of particular interest here is the *standard filtration* of a group algebra[3] by the powers of its augmentation ideal:

$$(1.3) \qquad KG \supset \underline{g} \supseteq \underline{g}^2 \supseteq \ldots .$$

The algebra S equipped with a filtration $\mathcal{S}$ becomes a *filtered algebra*. The corresponding *graded associative K-algebra*

$$T = \mathrm{gr}\, \mathcal{S}$$

is defined in the following way: as a K-module, it is the direct sum of the quotient K-modules

$$T_r = S^{(r)}/S^{(r+1)} \quad ;$$

and multiplication on T is specified by taking the product of the *homogeneous* elements $x + S^{(r+1)} \in T_r$ and $y + S^{(s+1)} \in T_s$ to be $xy + S^{(r+s+1)} \in T_{r+s}$. Thus

$$T_rT_s \subseteq T_{r+s} \quad (r, s \geq 0) .$$

In particular, the graded algebra corresponding to the standard filtration (1.3) of KG will be denoted by $\mathrm{gr}\, KG$.

If S, S' are K-algebras with filtrations $\mathcal{S}, \mathcal{S}'$, then a filtered homomorphism $f : S \to S'$ (mapping $S^{(r)}$ into $S'^{(r)}$ for all r) induces a graded homomorphism $\mathrm{gr}\, f : \mathrm{gr}\, \mathcal{S} \to \mathrm{gr}\, \mathcal{S}'$ (mapping the rth homogeneous component of $\mathrm{gr}\, \mathcal{S}$ into that of $\mathrm{gr}\, \mathcal{S}'$ for all r). Thus, $\mathrm{gr}(\cdot)$ is a functor from filtered K-algebras to graded K-algebras.

It is time to turn our attention to groups. Working still within the filtered

---

[2] More precisely, a non-negative integral filtration. No other filtrations will be considered here.

[3] We recall that the elements of the group algebra KG are the formal K-linear combinations

$$\sum_{g \in G} a_g g \quad (a_g \in K)$$

with a_g nonzero for only finitely many g , and that $\underline{g}$ consists of the elements with zero coefficient sum $\sum a_g$.

algebra S, let us assume that $1 \notin S^{(1)}$ and let G be a multiplicative subgroup of $1 + S^{(1)}$. Intuitively speaking, we are going to consider the "algebra of leading terms" of the elements of G.

Write

$$G^{(r)} = G \cap \left(1+S^{(r)}\right) \quad (r \geq 1) . \tag{1.4}$$

If $g = 1 + u \in G^{(r)}$, we call $u + S^{(r+1)} \in T_r$ the *leading term of* g *relative to* $G^{(r)}$ (it may of course be zero). The simple formulae

$$g^{-1} \equiv 1 - u \left(\text{mod } S^{(r+1)}\right) , \tag{1.5}$$

$$gg' \equiv 1 + u + u' \left(\text{mod } S^{(r+1)}\right) , \tag{1.6}$$

where also $g' = 1 + u' \in G^{(r)}$, show immediately that these leading terms relative to $G^{(r)}$ form an additive subgroup $\Lambda_r = \Lambda_r(G)$ of T_r. (Λ_r is *not* in general a K-submodule.) Next, if $g = 1 + u \in G^{(r)}$ and $h = 1 + v \in G^{(s)}$, then the identity

$$(g, h) = 1 + g^{-1}h^{-1}[u, v] ,$$

where

$$(g, h) = g^{-1}h^{-1}gh , \quad [u, v] = uv - vu ,$$

shows that

$$(g, h) \equiv 1 + [u, v] \left(\text{mod } S^{(r+s+1)}\right) , \tag{1.7}$$

whence

$$\left[\Lambda_r, \Lambda_s\right] \subseteq \Lambda_{r+s} \quad (r, s \geq 1) .$$

It follows that the direct sum $\Lambda = \Lambda(G)$ of the Λ_r is a *graded Lie subring* of gr S. So we have finally arrived at Lie rings!

If $g = 1 + u \in G^{(t)}$ but $g \notin G^{(t+1)}$, then the *nonzero* element $u + S^{(t+1)}$ of T_t is appropriately called the *leading term* of g. It is evident that $\Lambda(G)$ is the additive group generated by the leading terms of the elements of G and this gives perhaps the most intuitive definition. It should not be forgotten that all these definitions are relative to the given filtration S of S. When it is necessary to emphasize this fact, we shall write $\Lambda^{S}(G)$ for $\Lambda(G)$.

If H is a subgroup of G, then $\Lambda(H)$ is clearly a graded Lie subring of $\Lambda(G)$. It turns out that, if N is a normal subgroup of G, then $\Lambda(N)$ is in fact a graded Lie ideal of $\Lambda(G)$. The interpretation of the quotient graded Lie ring

$\Lambda(G)/\Lambda(N)$ will be given a little later.

As we shall now see, it is possible to construct an isomorphic copy Γ of Λ more directly in terms of G. Consider the series

(1.8) $$G : G = G^{(1)} \supseteq G^{(2)} \supseteq \dots .$$

By (1.5) and (1.6), each $G^{(r)}$ is a subgroup of G and, by (1.7),

(1.9) $$\left(G^{(r)}, G^{(s)}\right) \subseteq G^{(r+s)} \quad (r, s \geq 1) .$$

In virtue of these properties, the series (1.8) is called a *filtration*[4] of G. Every such filtration automatically satisfies $\left(G^{(r)}, G\right) \subseteq G^{(r+1)}$ and is therefore a central series. In particular, each $G^{(r)}$ is a normal subgroup of G and the quotient groups $\Gamma_r = G^{(r)}/G^{(r+1)}$ are abelian.

We observe now that, by (1.5) and (1.6), the mapping

$$\phi_r : \Gamma_r \to \Lambda_r ,$$

$$(1+u)G^{(r+1)} \mapsto u + S^{(r+1)} ,$$

is a group isomorphism. Taken all together, the ϕ_r induce an isomorphism of graded abelian groups $\phi : \Gamma \to \Lambda$, where Γ is the direct sum of the Γ_r. Finally, transferring the Lie multiplication on Λ back into Γ by means of ϕ^{-1}, we turn Γ into a graded Lie ring isomorphic to Λ. In view of (1.7), multiplication on Γ is specified by taking the product of the homogeneous elements $gG^{(r+1)} \in \Gamma_r$ and $hG^{(s+1)} \in \Gamma_s$ to be $(g, h)G^{(r+s+1)} \in \Gamma_{r+s}$. We denote Γ by gr G. In particular, when S is the standard filtration (1.3) of $S = KG$, we denote Γ by $\mathrm{gr}^K G$. In this case, $G^{(r)}$ is the rth *dimension subgroup* of G over K, denoted by $D_r(K, G)$.

The final step is to remove the ring-theoretical scaffolding: given *any* group filtration G - whether obtained from an algebra filtration or not - we may form the graded Lie ring gr G in the manner just described. Of course, it has to be proved that the construction actually works! But this can be done with the aid of the group-theoretical counterparts of the standard Lie identities, namely:

---

[4] Lazard [8] used the term *N-series*.

$$(g, g) = (g, h)(h, g) = 1,$$

$$(g, hk) = (g, k)(g, h)^k,$$

$$(g, h, k^g)(k, g, h^k)(h, k, g^h) = 1,$$

where $g^h = h^{-1}gh$, $(g, h, k) = ((g, h), k)$.

The most important filtration of a group which does not arise in an obvious way from an algebra filtration is the lower central series

(1.10) $$G = \gamma_1 G \supseteq \gamma_2 G \supseteq \dots ,$$

whose terms are defined inductively by

$$\gamma_{r+1} G = (\gamma_r G, G) \quad (r \geq 1).$$

Standard theorems of group theory show that this *is* a filtration and that

(1.11) $$\gamma_r G \subseteq G^{(r)} \quad (r \geq 1)$$

for every filtration (1.8) of G. The corresponding graded Lie ring is denoted by gr G. It is also quite easily proved that

(1.12) gr G *is generated (as a Lie ring) by its first homogeneous component* $\mathrm{gr}_1\, G$.

It is natural at this stage to pose the question:

Does every group filtration arise from a suitable ring filtration?

It is easy to show that, if the filtration (1.8) of G is induced by a ring filtration, then

$$G^{(r)} \supseteq D_r(\mathbb{Z}, G) \quad (r \geq 1).$$

On the other hand, by (1.11),

$$D_r(\mathbb{Z}, G) \supseteq \gamma_r G \quad (r \geq 1).$$

It follows that, *if the lower central series of G itself arises from a ring filtration,* then we must have

$$D_r(\mathbb{Z}, G) = \gamma_r G \quad (r \geq 1).$$

Although this last equation was long conjectured to be universally valid, Rips [16] eventually constructed a finite 2-group G for which $D_4(\mathbb{Z}, G) \neq \gamma_4 G$. The general answer to our question is therefore "no".

We record some simple formal properties of group filtrations. The assignment to each group filtration of its associated graded Lie ring becomes a functor when we define a morphism from one group filtration $\mathcal{U} : U = U^{(1)} \supseteq U^{(2)} \supseteq \dots$ to another $\mathcal{V} : V = V^{(1)} \subseteq V^{(2)} \subseteq \dots$ as a group homomorphism $\phi : U \to V$ such that $\phi\left(U^{(r)}\right) \subseteq V^{(r)}$ for all r. The induced homomorphism of graded Lie rings is denoted by $\operatorname{gr} \phi$.

Let $\mathcal{G}$ be the filtration (1.8) of G. Then its inverse image

$$\phi^{-1}(\mathcal{G}) : H = \phi^{-1}\left(G^{(1)}\right) \supseteq \phi^{-1}\left(G^{(2)}\right) \supseteq \dots$$

under a group homomorphism $\phi : H \to G$ is a filtration, ϕ is a morphism from $\phi^{-1}(\mathcal{G})$ to $\mathcal{G}$ and

$$\operatorname{gr} \phi : \operatorname{gr}\left(\phi^{-1}(\mathcal{G})\right) \to \operatorname{gr} \mathcal{G}$$

is injective. In particular, when H is a subgroup of G and ϕ the inclusion mapping, this yields the isomorphism

(1.13) $$\operatorname{gr}(\mathcal{G} \cap H) \cong (\operatorname{gr} \mathcal{G}) \cap H ,$$

where the rth member of $\mathcal{G} \cap H$ is $G^{(r)} \cap H$ and the rth component of $(\operatorname{gr} \mathcal{G}) \cap H$ is $\left(G^{(r)} \cap H\right)G^{(r+1)}/G^{(r+1)}$. (The notation in (1.13) is hybrid but useful.)

Again, the image

$$\psi(\mathcal{G}) : \psi(G) = \psi\left(G^{(1)}\right) \supseteq \psi\left(G^{(2)}\right) \supseteq \dots$$

of $\mathcal{G}$ under a group homomorphism $\psi : G \to F$ is a filtration of the image $\psi(G)$. If the sequence of group homomorphisms

$$1 \to N \xrightarrow{\phi} G \xrightarrow{\psi} F \to 1$$

is exact, then so is the sequence of graded Lie ring homomorphisms

$$0 \to \operatorname{gr}\left(\phi^{-1}(\mathcal{G})\right) \to \operatorname{gr} \mathcal{G} \to \operatorname{gr}\left(\psi(\mathcal{G})\right) \to 0 .$$

In particular, taking ψ as the canonical homomorphism of G onto a quotient group G/N, we get the isomorphism

(1.14) $$\operatorname{gr}(\mathcal{G}/N) \cong \operatorname{gr} \mathcal{G}/(\operatorname{gr} \mathcal{G}) \cap N ,$$

where

$$\mathcal{G}/N : G/N = G^{(1)}N/N \supseteq G^{(2)}N/N \supseteq \dots .$$

In the original situation where $\mathcal{G}$ was induced by an algebra filtration, the isomorphisms (1.13) and (1.14) can be expressed in the simple forms

(1.15) $$\operatorname{gr}(\mathcal{G} \cap H) \cong \Lambda(H) ,$$

(1.16) $$\operatorname{gr}(\mathcal{G}/N) \cong \Lambda(G)/\Lambda(N) .$$

In a similar way, $\operatorname{gr} \mathcal{G}$ defines a functor from groups to graded Lie rings. If

G is the lower central series of G, then G/N above is the lower central series of G/N and so (1.14) yields

(1.17) $$\mathrm{gr}(G/N) \cong \mathrm{gr}\, G/(\mathrm{gr}\, G) \cap N .$$

The ideas of this section are due to Magnus [9, 10, 11] and Lazard [8]. To put the matter briefly, Magnus introduced the "Λ" method and Lazard the "Γ" method.

2. An application

We illustrate the theory of §1 with a typical application to finite groups, due to Vaughan-Lee [18].

The *breadth* $b = b(G)$ of a finite p-group G is defined by

(2.1) $$p^b = \text{maximum size of a conjugacy class of } G .$$

Vaughan-Lee proved that

(2.2) $$|G'| \leq p^{\frac{1}{2}b(b+1)} ,$$

thereby verifying for p-groups a conjecture made by Wiegold [20] in the 1950's. Lie methods were first applied to the problem by P.M. Neumann [14], who proved the weaker result

$$|G'| \leq p^{b^2} .$$

The efficacy of Lie methods in a case like this depends on the close relation between the p-group G and the Lie ring

$$L = \mathrm{gr}\, G .$$

The product UV of additive subgroups U, V of L is defined as the additive subgroup generated by the Lie products uv $(u \in U, v \in V)$. The *lower central series* of L is

$$L \supseteq L^2 \supseteq L^3 \supseteq \dots ,$$

where the powers are defined inductively by $L^{r+1} = L^r L$. On the other hand, by its definition

$$L = L_1 \oplus \dots \oplus L_c ,$$

where $L_r = \gamma_r G/\gamma_{r+1} G$ and c is the nilpotency class of G. It is easy to see that

$$L^r = L_r \oplus \dots \oplus L_c \quad (r = 1, 2, \dots) ,$$

so that L is nilpotent of the same class c as G. We have

$$|L^r| = |\gamma_r G|$$

and, in particular

$$|L| = |G| \ , \quad |L^2| = |G'| \ .$$

Let $g_1, \ldots, g_d$ be a minimal set of generators of G. This is the same as saying that the corresponding elements

$$x_i = g_iG' \in G/G' = L_1$$

form a minimal set of generators of the additive group L_1. This in turn is equivalent to saying that

(2.3) $$x_1, \ldots, x_d$$

are a minimal set of generators of the Lie ring L.

The *breadth* $b(g)$ of $g \in G$ is defined by

$$p^{b(g)} = |\text{conjugacy class of } g| = |G|/|C(g)| \ ,$$

where $C(g) = \{h \in G : (g, h) = 1\}$. Clearly,

$$b(G) = \max_{g \in G} b(g) \ .$$

Similarly, the breadth $b(x)$ of $x \in L$ is defined by

$$p^{b(x)} = |Lx| = |L|/|C(x)| \ ,$$

where $C(x) = \{y \in L : xy = 0\}$.

LEMMA (P.M. Neumann). *If* $g \in G$ *and* $x = gG' \in L$, *then* $b(x) \le b(g)$.

Proof. Consider $C^*(g) = \text{gr } G \cap C(g) \subseteq L$, the notation being as in (1.13). If $h \in C(g) \cap \gamma_r G$, the element $y = h\gamma_{r+1}G$ of L_r is in $C^*(g)$; moreover, $C^*(g)$ is generated by such elements. Since $xy = (g, h)\gamma_{r+2}G = 0$, $C^*(g) \subseteq C(x)$. From the definition of $C^*(g)$, it is easily seen that $|C^*(g)| = |C(g)|$. Therefore

$$\begin{aligned} p^{b(x)} &= |L|/|C(x)| \le |L|/|C^*(g)| \\ &= |G|/|C(g)| = p^{b(g)} \ , \end{aligned}$$

which proves the lemma.

We pass now to the quotient Lie ring $\bar{L} = L/M$, where $M = pL + L^3$. This has the advantages of being a Lie algebra over $\mathbb{F}_p$ and being nilpotent of class 2 : $p\bar{L} = \bar{L}^3 = \{0\}$. As a vector space, $\bar{L} = \bar{L}_1 \oplus \bar{L}_2$, where

$$\bar{L}_i = (L_i+M)/M \cong L_i/pL_i \quad (i = 1, 2) \ .$$

Further, $\bar{L}_2 = \bar{L}^2$ and the canonical images

$$\bar{x}_1, \ldots, \bar{x}_d \in \bar{L}_1$$

of the elements (2.3) form a basis of $\bar{L}_1$; thus, the $\bar{x}_i$ form a minimal generating set for $\bar{L}$. The *breadth* $b(u)$ of $u \in \bar{L}$ is defined as the dimension of the subspace $\bar{L}u$ or equivalently as the codimension of $C(u)$. Since $C(u)$ properly contains the central ideal $\bar{L}^2$ (if $u \in \bar{L}^2$, $C(u) = \bar{L}$; if $u \notin \bar{L}^2$, $u \in C(u) - \bar{L}^2$), it follows that $b(u)$ is at most $d - 1$.

We are now in a position to get an upper bound for $|G'|$. First,

$$|G'| = |L^2| = |\bar{L}^2||N| , \tag{2.4}$$

where $N = pL^2 + L^3$. Since $\bar{L}^2$ is generated as a vector space by the Lie products $\bar{x}_i\bar{x}_j$ $(1 \le i < j \le d)$, we have

$$|\bar{L}^2| \le p^{\frac{1}{2}d(d-1)} . \tag{2.5}$$

Now it is an easy consequence of the Jacobi identity that $L^{r+1} = \sum_1^d L^r x_i$ $(r = 1, 2, \ldots)$. Therefore

$$N = \sum_i \left(pLx_i + L^2x_i\right) = \sum_i \left(Lx_i \cap N\right) ,$$

and so

$$|N| \le \prod_i |Lx_i \cap N| \le \prod_i |Lx_i \cap M| .$$

Since

$$Lx_i/\left(Lx_i \cap M\right) \cong \left(Lx_i + M\right)/M = \bar{L}\bar{x}_i ,$$

we get

$$|N| \le \prod_1^d \frac{|Lx_i|}{|\bar{L}\bar{x}_i|} . \tag{2.6}$$

Let us now *provisionally assume* that the x_i *may be chosen so that each* $\bar{x}_i$ *has the greatest possible breadth* $d - 1$. Then, in (2.6), $|Lx_i| \le p^b$ by the lemma and $|\bar{L}\bar{x}_i| = p^{d-1}$ by our assumption, so that

$$|N| \le p^{(b+1-d)d} . \tag{2.7}$$

Combining (2.4), (2.5) and (2.7), we get

$$|G'| \le p^{\frac{1}{2}(2b+1-d)d} . \tag{2.8}$$

However, d is an integer which, by (2.7), lies between 1 and $b+1$. The maximum of the right hand side of (2.8) for such d is $p^{\frac{1}{2}b(b+1)}$, attained when $d = b$ or $b+1$, and this proves the required inequality (2.2).

What happens when the provisional assumption is false? Vaughan-Lee circumvents this difficulty very cleverly by proving the following result.

LEMMA. *Let Λ be a Lie ring such that $p\Lambda = \Lambda^3 = \{0\}$, where p is a prime. Suppose Λ can be generated by d, but no fewer, elements and that Λ is not generated by its elements of breadth $d-1$. Then Λ has a proper Lie subring Δ such that $\Delta^2 = \Lambda^2$.*

In order to complete the proof of (2.2), choose a Lie subring D of $\bar{L}$ minimal subject to $D^2 = \bar{L}^2$ and carry out essentially the same proof as before with D in place of $\bar{L}$. We shall not go into the details.

Vaughan-Lee's Lemma is quite difficult to prove. It can be reformulated as a result in linear algebra as follows:

Let V be a finite dimensional vector space over $\mathbb{F}_p$ and let J be a subspace of the exterior square $V^{(2)}$ such that, for every $v \neq 0$ in V there exists $w \in V$ such that $v \wedge w$ is a nonzero element of J. Then V has a basis $v_1, \ldots, v_n$ such that

$$J + \langle v_1, \ldots, \hat{v}_i, \ldots, v_n \rangle^{(2)} = V^{(2)} \quad (i = 1, \ldots, n).$$

The assumption that the coefficient field is $\mathbb{F}_p$ seems to be irrelevant. It would be nice to have a simple direct proof of this straightforward looking result.

PART II

FORMAL POWER SERIES

Every free group can be embedded in a suitable algebra of formal power series, and such an algebra has a naturally defined filtration. Thus, the methods of §1 may be used to study the quotient groups of free groups. The coefficient ring for the formal power series can be chosen to suit the needs of the particular problem under consideration.

3. Magnus algebras

We begin with the free associative K-algebra A on a set X. When it is necessary to indicate the dependence of A on X, or on K and X, we shall write

$A(X)$ or $A(K, X)$. A similar convention will apply to the various structures associated with A.

We may think of the elements of X as non-commuting "variables" and the elements of A as polynomials in these variables. Let A_r denote the K-submodule of A (freely) generated by the monomials $x_1 \dots x_r$ $(x_i \in X)$; in other words, the elements of A_r are the homogeneous polynomials of degree r. Then, as a K-module,

(3.1) $$A = \bigoplus_0^\infty A_r$$

and we have

$$A_r A_s \subseteq A_{r+s} \quad (r, s \geq 0) ,$$

so that A is a graded algebra.

Similarly, if L is the Lie K-subalgebra of A generated by X, then it is easy to see that

$$L = \bigoplus_1^\infty L_r ,$$

$$[L_r, L_s] \subseteq L_{r+s} \quad (r, s \geq 1) ,$$

where $L_r = L \cap A_r$. Thus, L is a graded Lie subalgebra of A. The following properties of L are analogous to properties of A but lie deeper[5]:

(3.2) *L is the free Lie K-algebra on X ;*

(3.3) *both L_r and A_r/L_r are free K-modules* $(r \geq 1)$.

If k is a subring of K, we may regard $A(k, X)$ as a k-subalgebra of $A(K, X)$ in the obvious way. It is a consequence of (3.3) that then

(3.4) $$L(k, X) = L(K, X) \cap A(k, X) .$$

An endomorphism ϕ of A is said to be *graded* when $\phi(A_r) \subseteq A_r$ for all r. Such a graded endomorphism clearly maps L into itself and therefore induces a graded endomorphism of L ; that is, $\phi(L_r) \subseteq L_r$ for all r. Now, the K-linear mappings $\theta : A_1 \to A_1$ form a monoid Ω under multiplication and, since A is the free algebra on X, every element of Ω can be extended uniquely to a graded endomorphism of A. Thus, each A_r and L_r is in a natural way a $K\Omega$-module.

Let us now extend the K-module in (3.1) to the product K-module

---

[5] They are consequences of the Poincaré-Birkhoff-Witt Theorem.

(3.5) $$\hat{A} = \prod_{0}^{\infty} A_r .$$

We may regard the elements of $\hat{A}$ as *formal power series*

(3.6) $$f = f_0 + f_1 + \dots \quad (f_r \in A_r)$$

and the K-module structure is given by

$$(\lambda f + \mu g)_r = \lambda f_r + \mu g_r \quad (\lambda, \mu \in K) .$$

We define multiplication on $\hat{A}$ by

$$(fg)_r = \sum_{0}^{r} f_i g_{r-i} .$$

These are the formal algebraic operations one would expect from the name "formal power series" and $\hat{A}$ becomes an associative K-algebra containing A as subalgebra. We call $\hat{A}$ the *Magnus K-algebra on X*. The series (3.6) is called a *Lie series* when each component f_r is in L. The Lie series form a Lie K-subalgebra $\hat{L}$ of $\hat{A}$.

If $f_0 = \dots = f_{r-1} = 0$ but $f_r \neq 0$, the power series (3.6) is said to have *order* r; 0 has order $+\infty$. The power series of order greater than or equal to r form an ideal $\hat{\underline{a}}^{(r)}$ of $\hat{A}$ and we have the filtration

(3.7) $$\mathrm{A} : \hat{A} = \hat{\underline{a}}^{(0)} \supseteq \hat{\underline{a}}^{(1)} \supseteq \dots .$$

It is natural to ask whether $\hat{\underline{a}}^{(r)}$ is the rth power of $\hat{\underline{a}} = \hat{\underline{a}}^{(1)}$. For fixed $r > 1$, this is true when X is finite but false when X is infinite - a result which is not quite easy to prove.

Since

$$A_r + \hat{\underline{a}}^{(r+1)} = \hat{\underline{a}}^{(r)} , \quad A_r \cap \hat{\underline{a}}^{(r+1)} = \{0\} ,$$

it follows that $\mathrm{gr}\,\mathrm{A}$ is isomorphic, as graded algebra, to A. Henceforth *we shall identify* $\mathrm{gr}\,\mathrm{A}$ *with* A.

An endomorphism ϕ of $\hat{A}$ is said to be *filtered* if $\phi\left(\hat{\underline{a}}^{(r)}\right) \subseteq \hat{\underline{a}}^{(r)}$ for all r. Such a filtered endomorphism induces a graded endomorphism $\mathrm{gr}\,\phi$ of $\mathrm{gr}\,\mathrm{A} = A$. The relationship of $\mathrm{gr}\,\phi$ to ϕ is very simple: if $x \in X$ and $\phi(x) = \xi_1 + \xi_2 + \dots \quad (\xi_r \in A_r)$, then $(\mathrm{gr}\,\phi)(x) = \xi_1$. Clearly,

$$\mathrm{gr}(\phi\phi') = (\mathrm{gr}\,\phi)(\mathrm{gr}\,\phi') ,$$

and, if ϕ is invertible,

$$\mathrm{gr}\left(\phi^{-1}\right) = (\mathrm{gr}\,\phi)^{-1} .$$

FORMAL IMPLICIT FUNCTION THEOREM. *A filtered endomorphism* ϕ *of* $\hat{A}$ *is invertible if, and only if,* $\mathrm{gr}\ \phi$ *is invertible.*

$\hat{A}$ has a completeness property which can be expressed in most concrete form as follows: under the distance function

$$d(f, g) = 2^{-\mathrm{ord}(f-g)} ,$$

$\hat{A}$ is a *complete* metric space. The (open and closed) sets $f + \hat{\underline{\underline{a}}}^{(r)}$ $(r \geq 0)$ form a fundamental system of neighbourhoods of f, for $f + \hat{\underline{\underline{a}}}^{(r)}$ is just the closed ball of radius 2^{-r} and centre f. Further, $\hat{A}, \hat{L}$ are the closures of A, L, and $\hat{\underline{\underline{a}}}^{(r)}$ is the closure of $\underline{\underline{a}}^{r}$, where $\underline{\underline{a}} = A \cap \hat{\underline{\underline{a}}}$.

Convergence and continuity in the sense of the metric are called *formal convergence* and *formal continuity*. We note the simple result that an endomorphism ϕ of $\hat{A}$ is filtered if, and only if, it maps $\hat{\underline{\underline{a}}}$ into itself and is formally continuous. Let us see what is entailed in the formal convergence of an infinite series

$$f^{(1)} + f^{(2)} + \dots$$

whose terms $f^{(r)}$ are themselves formal power series. The Cauchy condition comes down to

$$\lim_{r\to\infty} \mathrm{ord}\ f^{(r)} = +\infty ,$$

and by completeness this implies formal convergence. The rth homogeneous component of the sum is just the rth homogeneous component of the *finite* sum $f^{(1)} + \dots + f^{(n)}$, provided that n is chosen large enough to ensure that $\mathrm{ord}\ f^{(s)} > r$ whenever $s > n$. For example, the series

$$\lambda_0 + \lambda_1 v + \lambda_2 v^2 + \dots \quad (\lambda_i \in K)$$

is formally convergent whenever $v \in \hat{\underline{\underline{a}}}$. As a further illustration, we have the following result:

(3.8) *every mapping of* X *into* $\hat{\underline{\underline{a}}}$ *can be uniquely extended to a filtered endomorphism of* $\hat{A}$.

Let us briefly indicate the proof. Since A is the free algebra on X, the given mapping can be extended uniquely to a homomorphism $\psi : A \to \hat{A}$. Next we extend ψ to $\hat{\psi} : \hat{A} \to \hat{A}$ "by continuity"; that is, if

$$f = f_0 + f_1 + \dots \quad (f_r \in A_r) ,$$

then

$$(*) \qquad \hat{\psi}(f) = \psi(f_0) + \psi(f_1) + \dots .$$

Since $\psi(\underline{\underline{a}}^r) \subseteq \hat{\underline{\underline{a}}}^r \subseteq \hat{\underline{\underline{a}}}^{(r)}$, the series in (*) is formally convergent. It is now a routine matter to verify that the mapping $\hat{\psi}$ so defined is indeed the unique filtered endomorphism extending ψ.

One of the main reasons for working with the Magnus algebra is the fact that $1 + \hat{\underline{\underline{a}}}$ *is a multiplicative subgroup of* $\hat{A}$: for, if $u \in \hat{\underline{\underline{a}}}$, $1 + u$ has the inverse

$$(1+u)^{-1} = 1 - u + u^2 - \dots .$$

Let G be a subgroup of $1 + \hat{\underline{\underline{a}}}$. Remembering that gr $\hat{A}$ is identified with A, we see that $\Lambda(G) = \Lambda^{\hat{A}}(G)$ is determined as follows. Write $g \in G$ in the form

$$g = 1 + u_r + u_{r+1} + \dots ,$$

where $u_i \in A_i$, $u_r \neq 0$. Then u_r is the *leading term* of g, and $\Lambda(G)$ is simply the additive subgroup of A generated by the leading terms of the elements of G. We note the following result:

(3.9) *Let* G *be a subgroup of* $1 + \hat{\underline{\underline{a}}}$. *If* ψ *is a filtered endomorphism of* $\hat{A}$, *then*

$$(\text{gr } \psi)\Lambda(G) \subseteq \Lambda(\psi G) .$$

Proof. It will be sufficient to prove that, if u_r is the leading term of $g = 1 + u_r + u_{r+1} + \dots \in G$, then $(\text{gr } \psi)u_r$, if nonzero, is the leading term of $\psi g = 1 + \psi u_r + \psi u_{r+1} + \dots$. But this is clear by inspection.

MAGNUS' EMBEDDING THEOREM[6]. *Let* F *be the subgroup of* $1 + \hat{\underline{\underline{a}}}$ *generated by* $1 + X$. *Then* F *is freely generated by* $1 + X$ *and its elements are linearly independent over* K.

Proof. We consider words

$$w = (1+x_1)^{m_1} \dots (1+x_r)^{m_r} ,$$

where $x_i \in X$, $m_i \neq 0$ and adjacent x_i are distinct. Then r is the *width* of w and $(x_1, \dots, x_r)$ the *associated sequence*. It is sufficient to prove that

$$\sum \lambda_w w = 0 \Rightarrow \text{all } \lambda_w = 0 ,$$

where the summation is over some finite set W of such words w.

---

6 The first statement is due to Magnus [9] and the second is implicit in Fox [3].

Consider those w in W associated with one particular sequence $(x_1, \ldots, x_r)$ of *maximum* width. The contribution from these w is say,

$$(*) \qquad \sum \lambda_{m_1,\ldots,m_r} (1+x_1)^{m_1} \ldots (1+x_r)^{m_r} .$$

It will be sufficient to show that all $\lambda_{m_1,\ldots,m_r}$ are zero.

Now, the coefficient of any monomial $x_1^{t_1} \ldots x_r^{t_r}$ with all $t_i \neq 0$ in the expansion of (*) is the same as in the expansion of the original sum $\sum \lambda_w w$. It follows that

$$\sum \lambda_{m_1,\ldots,m_r} \left((1+x_1)^{m_1}-1\right) \ldots \left((1+x_r)^{m_r}-1\right) = 0 .$$

However, this implies that

$$\sum \lambda_{m_1,\ldots,m_r} \left((1+\xi_1)^{m_1}-1\right) \ldots \left((1+\xi_r)^{m_r}-1\right) = 0$$

in the formal power series ring $K[[\xi_1, \ldots, \xi_r]]$, where the ξ_i are *distinct*, *commuting* variables. It is therefore sufficient to prove the linear independence of the elements $\left((1+\xi_1)^{m_1}-1\right) \ldots \left((1+\xi_r)^{m_r}-1\right)$ with all $m_i \neq 0$ in $K[[\xi_1, \ldots, \xi_r]]$. Then, by induction on r, it suffices to prove the linear independence of the elements $(1+\xi_r)^{m_r} - 1$ $(m_r \neq 0)$ over $K[[\xi_1, \ldots, \xi_{r-1}]]$. But that is easy!

The theorem allows us to identify the group algebra KF with the K-submodule of $\hat{A}$ generated by F. Let $\underline{f}$ denote the augmentation ideal of F. Consider the filtration of F induced by the filtration A of $\hat{A}$, namely,

$$(3.10) \qquad \mathrm{F} : F = F^{(1)} \supseteq F^{(2)} \supseteq \ldots ,$$

where

$$F^{(r)} = F \cap \left(1+\hat{\underline{\underline{a}}}^{(r)}\right).$$

We show now that

$$(3.11) \qquad F^{(r)} = F \cap \left(1+\underline{\underline{f}}^r\right) = D_r(K, F) .$$

This yields

$$(3.12) \qquad \Lambda(F) \cong \mathrm{gr}\ \mathrm{F} \cong \mathrm{gr}^K F ,$$

in the notation introduced in §1.

Since $A \subseteq KF \subseteq \hat{A}$ and $\underline{\underline{a}}^r \subseteq \underline{\underline{f}}^r \subseteq \hat{\underline{\underline{a}}}^{(r)}$, where $\underline{\underline{a}} = A \cap \hat{\underline{\underline{a}}}$, we have homomorphisms

$$A/\underline{\underline{a}}^r \xrightarrow{\phi} KF/\underline{\underline{f}}^r \xrightarrow{\psi} \hat{A}/\hat{\underline{\underline{a}}}^{(r)} .$$

It is easy to see that $A + \hat{\underline{\underline{a}}}^{(r)} = \hat{A}$ and $A \cap \hat{\underline{\underline{a}}}^{(r)} = \underline{\underline{a}}^r$, whence $\psi \circ \phi$ is an isomorphism. Hence ϕ is injective. However, the identity

$$(1+x)^{-1} = -(1+x)^{-1}x^r + \sum_0^{r-1} (-x)^i$$

shows that $KF = A + \underline{\underline{f}}^r$, so that ϕ is also surjective. Thus, both ϕ and ψ are isomorphisms and so we have

(3.13) $$A/\underline{\underline{a}}^r \cong KF/\underline{\underline{f}}^r \cong \hat{A}/\hat{\underline{\underline{a}}}^{(r)} .$$

In particular, since ψ is injective,

$$KF \cap \hat{\underline{\underline{a}}}^{(r)} = \underline{\underline{f}}^r ,$$

which implies the required result (3.11).

The originator of the theory we have outlined was Wilhelm Magnus, who was concerned in the first instance with power series over the integers. The following result was conjectured by Magnus [9] and later proved independently by Magnus [10] and Witt [21].

MAGNUS-WITT THEOREM. *If* $K = \mathbb{Z}$,

(3.14) $$\text{gr } F = \text{gr}^{\mathbb{Z}} F \cong \Lambda(F) = L .$$

Proof. We first prove the last statement, namely,

(3.15) $$\Lambda\big(F(\mathbb{Z}, X)\big) = L(\mathbb{Z}, X) .$$

It is sufficient to prove that

$$\Lambda\big(F(\mathbb{Z}, X)\big) \subseteq L(\mathbb{Z}, X)$$

because the reverse inclusion is obvious. Let us embed $\hat{A}(\mathbb{Z}, X)$ in $\hat{A}(\mathbb{Q}, X)$. Then $\Lambda\big(F(\mathbb{Z}, X)\big) = \Lambda\big(F(\mathbb{Q}, X)\big)$ and by (3.4), $L(\mathbb{Z}, X) = L(\mathbb{Q}, X) \cap \hat{A}(\mathbb{Z}, X)$. It is therefore sufficient to prove that

$$\Lambda\big(F(\mathbb{Q}, X)\big) \subseteq L(\mathbb{Q}, X) .$$

For $u \in \hat{\underline{\underline{a}}}(\mathbb{Q}, X)$, we write

$$e^u = \sum_0^\infty \frac{u^n}{n!} .$$

By (3.8), there is a unique filtered endomorphism ϕ of $\hat{A}(\mathbb{Q}, X)$ such that

$$\phi(x) = e^x - 1 \quad (x \in X) .$$

Clearly, $\operatorname{gr}\phi$ is the identity. By the Formal Implicit Function Theorem, ϕ is an automorphism and by (3.9),

$$\Lambda\{F(\mathbb{Q}, X)\} = \Lambda\{\tilde{F}(\mathbb{Q}, X)\} ,$$

where

$$\tilde{F}(\mathbb{Q}, X) = \phi\{F(\mathbb{Q}, X)\}$$

$$= \text{group generated by the elements } e^x \quad (x \in X) .$$

Thus, it is sufficient to prove that

$$\Lambda\{\tilde{F}(\mathbb{Q}, X)\} \subseteq L(\mathbb{Q}, X) . \tag{3.16}$$

Here we appeal to the *Baker-Campbell-Hausdorff Theorem*, which asserts that

$$\{e^u : u \in \hat{L}(\mathbb{Q}, X)\}$$

is a multiplicative subgroup of $\hat{A}(\mathbb{Q}, X)$.

It follows that every element of $\tilde{F}(\mathbb{Q}, X)$ has the form e^u , where $u \in \hat{L}(\mathbb{Q}, X)$ The leading term of such an element is the same as the leading term of $1 + u$, which is evidently in $L(\mathbb{Q}, X)$. This proves (3.16) and hence (3.15).

After our excursion into $\hat{A}(\mathbb{Q}, X)$, we return to $\hat{A}(\mathbb{Z}, X)$, using the simpler notation $\hat{A}, F, L$. The proof of the remaining part of (3.14) is independent of the part already established and rests on the fact that L is the free Lie ring on X . Since

$$\gamma_r F \subseteq F^{(r)} \quad (r \geq 1) ,$$

we get a homomorphism $\operatorname{gr} F \to \operatorname{gr}^{\mathbb{Z}} F$. By (3.12), we have an isomorphism $\operatorname{gr}^{\mathbb{Z}} F \to \Lambda(F)$. Composing these, we get a homomorphism $\theta : \operatorname{gr} F \to \Lambda(F)$ with $\theta\{(1+x)F'\} = x$ $(x \in X)$. Now, by (1.12), the elements $(1+x)F'$ generate $\operatorname{gr} F$ and so $\theta(\operatorname{gr} F) = L$. Thus, θ induces a homomorphism

$$\psi : \operatorname{gr} F \to L$$

such that $\psi\{(1+x)F'\} = x$ $(x \in X)$. On the other hand, since L is the free Lie ring on X , there is a homomorphism

$$\chi : L \to \operatorname{gr} F$$

such that $\chi(x) = (1+x)F'$ $(x \in X)$. Since $\chi \circ \psi$ fixes each generator $(1+x)F'$ of $\operatorname{gr} F$ and $\psi \circ \chi$ fixes each generator x of L , it follows that χ, ψ are mutually inverse isomorphisms. This completes the proof of the theorem.

For coefficient rings K other than $\mathbb{Z}$, $\operatorname{gr}^K F$ is in general different from $\operatorname{gr} F$, and $\Lambda(F)$ from L . We note that the first part of (3.14) is equivalent to

$$D_r(\mathbb{Z}, F) = \gamma_r F \quad (r \geq 1) . \tag{3.17}$$

4. Varieties of groups

The properties of a variety[7] may be studied through its free groups, each of which is a quotient of an absolutely free group by a fully invariant subgroup. The theory developed in §3 can be brought into play by embedding the absolutely free group in question in an appropriate Magnus algebra. In the present section, we touch on several matters related to varieties.

Let $\underline{\underline{G}}$ be the category of all groups and $\underline{\underline{C}}$ a full subcategory; in the sequel, $\underline{\underline{C}}$ will be either $\underline{\underline{G}}$ itself or the category, $\underline{\underline{F}}$, of all absolutely free groups. A functor $\theta : \underline{\underline{C}} \to \underline{\underline{G}}$ will be called a *subgroup functor* if

(a) for every object G in $\underline{\underline{C}}$, $\theta(G)$ is a subgroup of G, and

(b) for every morphism $f : G \to H$ in $\underline{\underline{C}}$, the following diagram commutes:

$$\begin{array}{ccc} G & \xrightarrow{\quad f \quad} & H \\ \uparrow & & \uparrow \\ \theta(G) & \xrightarrow{\quad \theta(f) \quad} & \theta(H) \end{array} .$$

Taking $G = H$ in the diagram and remembering that $\underline{\underline{C}}$ is a full subcategory of $\underline{\underline{G}}$, we see that $\theta(G)$ is always a fully invariant subgroup of G.

EXAMPLE 1. $\theta(G) = D_r(K, G)$ (for fixed r and K).

EXAMPLE 2. Let $\underline{\underline{V}}$ be a variety of groups. It is well known that, for each group G, there is a fully invariant subgroup $v(G) = v_{\underline{\underline{V}}}(G)$ such that a quotient group G/N lies in $\underline{\underline{V}}$ if, and only if, $N \supseteq v(G)$. Then $v : \underline{\underline{C}} \to \underline{\underline{G}}$ is a subgroup functor. Such subgroup functors may be called *varietal*. If $\underline{\underline{C}}$ is $\underline{\underline{F}}$ or $\underline{\underline{G}}$, $\underline{\underline{V}}$ is uniquely determined by v; indeed, $\underline{\underline{V}}$ is determined by $v(G)$, where G is any free group of infinite rank.

LEMMA 1. *Let ω be a functor from sets to groups. If $\lambda : X \to Y$ is surjective, so is $\omega(\lambda) : \omega(X) \to \omega(Y)$.*

Proof. λ has a right inverse μ and so $\omega(\lambda)$ has a right inverse $\omega(\mu)$.

LEMMA 2. *Every subgroup functor $\theta : \underline{\underline{F}} \to \underline{\underline{G}}$ is varietal.*

Proof. Let X be any infinite set and let $\Phi(Y)$ denote the free group on a set Y. Since $\theta\big(\Phi(X)\big)$ is fully invariant in $\Phi(X)$, there is a unique variety $\underline{\underline{V}}(X)$ such that

$$\theta\big(\Phi(X)\big) = v_{\underline{\underline{V}}(X)}\big(\Phi(X)\big) .$$

---

[7] For the basic facts about varieties of groups, see Hanna Neumann [13].

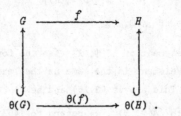

If now Y is any set such that $|Y| \le |X|$, then there exists a surjection $\lambda : X \to Y$. Since both $\theta\{\Phi(\cdot)\}$ and $v_{\underline{\underline{V}}(X)}\{\Phi(\cdot)\}$ are functors from sets to groups, it follows from Lemma 1 that

$$\theta\{\Phi(Y)\} = v_{\underline{\underline{V}}(X)}\{\Phi(Y)\} .$$

In particular, when Y is infinite this implies that $\underline{\underline{V}}(X) = \underline{\underline{V}}(Y)$, for $\underline{\underline{V}}(Y)$ is the *unique* variety such that

$$\theta\{\Phi(Y)\} = v_{\underline{\underline{V}}(Y)}\{\Phi(Y)\} .$$

It is now clear how to complete the proof and the details will be omitted.

COROLLARY. *For fixed* r *and* K *, there is a unique variety* $\underline{\underline{V}}$ *such that*

(4.1) $$v_{\underline{\underline{V}}}(F) = D_r(K, F)$$

for all free groups F *.*

NOTATION. If $\underline{\underline{V}}$ is the variety in the corollary, then, for an *arbitrary* group G , we write

(4.2) $$v_{\underline{\underline{V}}}(G) = \Delta_r(K, G) .$$

LEMMA 3. *A subgroup functor* $\theta : \underline{G} \to \underline{G}$ *is varietal if, and only if,* $\theta(f)$ *is surjective whenever* f *is.*

The proof, which is a simple extension of that of Lemma 2, will be omitted.

COROLLARY. $\Delta_r(K, G) \subseteq D_r(K, G)$ *for all groups* G .

The question of the precise relation between $\Delta_r(K, G)$ and $D_r(K, G)$ is a delicate one, as the following example indicates.

EXAMPLE. If $K = \mathbb{Z}$, then, by (3.17), the variety $\underline{\underline{V}}$ in (4.1) is the variety of all nilpotent groups of class less than r and so

(4.3) $$\Delta_r(\mathbb{Z}, G) = \gamma_r G$$

for all groups G . The example of Rips cited in §1 shows that there are groups G for which $\Delta_r(\mathbb{Z}, G) \ne D_r(\mathbb{Z}, G)$.

Let us return to the Magnus algebra $\hat{A} = \hat{A}(K, X)$ of §3. We recall that the natural filtration of $\hat{A}$ induces the filtration

$$\mathsf{F} : F = F^{(1)} \supseteq F^{(2)} \supseteq \dots$$

of the free subgroup F and that, if N is a normal subgroup of F , then

$$\mathrm{gr}(\mathsf{F}/N) \cong \Lambda(F)/\Lambda(N) ,$$

where

$$\mathsf{F}/N \,:\, F/N = F^{(1)}N/N \supseteq F^{(2)}N/N \supseteq \ldots \,.$$

Since $F^{(r)} = D_r(K, F) = \Delta_r(K, F)$, it follows that

$$F^{(r)}N/N = \Delta_r(K, F/N) \quad (r \geq 1) \,. \tag{4.4}$$

This result shows that, given K , the filtration F/N depends only on the group F/N and not on its presentation; it is therefore permissible to introduce the notation

$$\Lambda(F/N) = \Lambda(F)/\Lambda(N) \,. \tag{4.5}$$

A second consequence of (4.3) is that

$$G = \Delta_1(K, G) \supseteq \Delta_2(K, G) \supseteq \ldots$$

is a filtration for every group G .

Equation (4.5) expresses the essential point in Magnus' programme for applying Lie methods to study groups: one represents the given group as a quotient F/N and thereby associates with it the Lie ring $\Lambda(F)/\Lambda(N)$. Although this procedure lacks the conceptual simplicity of Lazard's method described in §1, it has the practical advantage of bringing formal power series directly into play. Magnus in fact concentrated on the case $K = \mathbb{Z}$, where $\Lambda(F/N) = \mathrm{gr}(F/N)$ by (4.3), (4.4).

The relation between D_r and Δ_r can be described explicitly as follows. The canonical algebra homomorphism

$$\phi : \hat{A} \to \hat{A}/\hat{\underline{\mathrm{a}}}^{(r)}$$

induces the canonical group homomorphism

$$F \to F/F^{(r)}$$

if we make the obvious identification of $F/F^{(r)}$ with $\phi(F)$. Suppose

$$\Delta_r(K, F/N) = S/N \,, \quad D_r(K, F/N) = T/N \,.$$

Then

$$S/F^{(r)} = \phi(N) \,,$$

$$T/F^{(r)} = \phi(F) \cap (1+J) \,,$$

where J is the ideal of $\hat{A}/\hat{\underline{\mathrm{a}}}^{(r)}$ generated by the elements $u - 1$, $u \in \phi(N)$.

Let $\underline{\underline{\mathrm{V}}}$ be a variety and write

$$v(G) = v_{\underline{\underline{\mathrm{V}}}}(G) \,.$$

Then

(4.6) $$F_{\underline{V}}(X) = F_{\underline{V}} = F/v(F)$$

is essentially the *free group of* $\underline{V}$ *on the set* $1 + X$. Indeed, the set

$$\{(1+x)v(F) : x \in X\}$$

generates $F_{\underline{V}}$ and every mapping of this set into a group G in $\underline{V}$ can be uniquely extended to a homomorphism of $F_{\underline{V}}$ into G. The variety $\underline{V}$ is determined by the particular group

(4.7) $$F_{\underline{V}}(X_\infty) = F(X_\infty)/v\big(F(X_\infty)\big) ,$$

where X_∞ is a countably infinite set. In this connexion, the elements of $F(X_\infty)$ are often referred to as *words* and those of $v\big(F(X_\infty)\big)$ as the *identical relators* for $\underline{V}$.

The Lie counterparts of (4.6) and (4.7) are $\Lambda\big(F_{\underline{V}}(X)\big)$ and

(4.8) $$\Lambda\big(F_{\underline{V}}(X_\infty)\big) = \Lambda\big(F(X_\infty)\big)/\Lambda\big(v\big(F(X_\infty)\big)\big) .$$

We may think of the elements of $\Lambda\big(F(X_\infty)\big)$ as *Lie words* (relative to K) and those of $\Lambda\big(v\big(F(X_\infty)\big)\big)$ as *identical Lie relators* for $\underline{V}$ (relative to K).

If X is a subset of Y, and if $A(X)$ is regarded as a subalgebra of $A(Y)$ in the usual way, then we have

(4.9) $$\Lambda\big(F_{\underline{V}}(X)\big) \cong \big[\Lambda\big(F(X)\big)+\Lambda\big(v\big(F(Y)\big)\big)\big]/\Lambda\big(v\big(F(Y)\big)\big)$$

or equivalently

(4.10) $$\Lambda\big(v\big(F(X)\big)\big) = A(X) \cap \Lambda\big(v\big(F(Y)\big)\big) .$$

Indeed, (4.9) describes the embedding of $\Lambda\big(F_{\underline{V}}(X)\big)$ in $\Lambda\big(F_{\underline{V}}(Y)\big)$ which we obtain by applying the functor $\Lambda\big(F_{\underline{V}}(\cdot)\big)$ to the commutative diagram

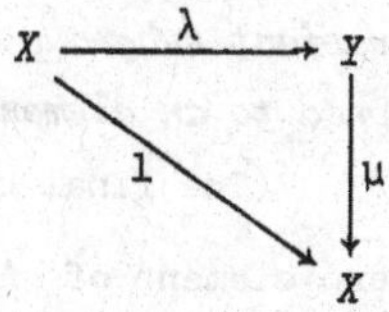

where λ is the embedding of X in Y and μ a left inverse[8] of λ. Intuitively, (4.9) and (4.10) tell us that, when dealing with identical Lie relators in a certain set of variables, we may, without harm, adjoin such further variables as are convenient.

If N is fully invariant in F, is $\Lambda(N)$ in any sense fully invariant in $\Lambda(F)$? The next two results are concerned with this question.

---

[8] This argument fails when X is empty - but (4.9) is obvious in this case.

(4.11) *If* N *is a fully invariant subgroup of* F , $\Lambda(N)$ *is mapped into itself by every graded endomorphism* θ *of* A *which maps* $\Lambda(F)$ *into itself.*

(Note that the final clause is redundant when $K = \mathbb{Z}$.)

Proof. The first homogeneous component $\Lambda_1(F)$ of $\Lambda(F)$ is just the additive group generated by X . Therefore θ maps each $x \in X$ into an element $m_1x_1 + \ldots + m_tx_t$, where $x_i \in X$, $m_i \in \mathbb{Z}$. Let ψ be the endomorphism of F which maps each $1+x \in 1+X$ to the corresponding product $(1+x_1)^{m_1} \ldots (1+x_t)^{m_t}$. By (3.8), we may extend ψ uniquely to a filtered endomorphism ϕ of $\hat{A}$, and from the definition of ψ and ϕ we see that $\theta = \text{gr}\,\phi$. By (3.9), $\theta(\Lambda(N)) = (\text{gr}\,\phi)(\Lambda(N)) \subseteq \Lambda(N)$. This completes the proof.

In §3 we drew attention to the fact that each A_r and L_r is a $K\Omega$-module, where Ω is the multiplicative monoid of all K-linear mappings of A_1 into itself. Let Ω^0 denote the submonoid of Ω formed by those K-linear mappings of A_1 into itself which map each $x \in X$ into an integral linear combination of elements of X . Then (4.11) and its proof show that $\Lambda_r(F)$ is a $\mathbb{Z}\Omega^0$-module and that, if N is fully invariant in F , $\Lambda_r(N)$ is a $\mathbb{Z}\Omega^0$-submodule.

Let $x_1, \ldots, x_r$ be distinct elements of X . Then

$$u = \sum_{\pi} \alpha_\pi x_{\pi 1} \ldots x_{\pi r} \quad (\alpha_\pi \in K) ,$$

where summation is over the permutations π of $1, \ldots, r$, is called a *multilinear element* of A .

(4.12) *Let* N *be a fully invariant subgroup of* F *and* u *a multilinear element of* $\Lambda(N)$. *Then* u *is mapped to an element of* $\Lambda(N)$ *by every endomorphism of* A *which maps* $\Lambda(F)$ *into itself.* (The final clause *is* needed even when $K = \mathbb{Z}$.)

Proof. With u above a nonzero element of $\Lambda(N)$, we have to prove that

$$v = \sum_{\pi} \alpha_\pi y_{\pi 1} \ldots y_{\pi r} \in \Lambda(N)$$

whenever $y_1, \ldots, y_r$ are nonzero elements of $\Lambda(F)$. We may assume the y_i homogeneous. By the definitions of $\Lambda(F)$ and $\Lambda(N)$, we may choose $h_1, \ldots, h_r \in F$ with leading terms $y_1, \ldots, y_r$ and

$$g = 1 + u_r + u_{r+1} + \dots \in N ,$$

where $u_r = u$, $u_i \in A_i$. If ϕ is an endomorphism of F with $\phi(1+x_i) = h_i$ $(1 \le i \le r)$, then

$$\phi(g) = 1 + \phi'(u_r) + \phi'(u_{r+1}) + \dots ,$$

where ϕ' is the filtered endomorphism of $\hat{A}$ extending ϕ . It is clear that v is the leading term of $1 + \phi'(u_r)$. We now state without proof of the following

LEMMA. *Let* s *be a positive integer. Then it is possible to choose* g *so that* $\sum_0^s u_{r+i}$ *is a linear combination of monomials* $z_1 z_2 \dots$, *where* $z_i \in X$, $\{x_1, \dots, x_r\} \subseteq \{z_1, z_2, \dots\}$.

If we take s as the sum of the degrees of the y_i and choose g as in the lemma, then v *is* the leading term of $\phi(g)$. Since $\phi(g) \in N$, $v \in \Lambda(N)$ as required.

PART III

p-FILTRATIONS

The general theory developed in earlier sections has some interesting special features when the characteristic of the coefficient ring K is a prime p . Following an account of these, we shall indicate some applications to finite groups of exponent p .

5. Filtrations of rings and groups (characteristic p version)

Formal identities in characteristic p are often conveniently expressed within a free algebra

$$A = A(\mathsf{F}_p, \{x_1, x_2, \dots\}) .$$

We quote two such identities, namely,

$$(x_1 + x_2 + \dots)^p - x_1^p - x_2^p - \dots = l(x_1, x_2, \dots) \in L , \tag{5.1}$$

$$\left[x_1, x_2^p\right] = [\underbrace{x_1, x_2, \dots, x_2}_{p \text{ terms}}] . \tag{5.2}$$

(5.1) is usually referred to as *Jacobson's identity*; notice that the identity in an arbitrary number of variables is an obvious consequence of the two variable case.

We return now to the situation of §1, using the original notation but assuming that K has prime characteristic p. We briefly recall the main points of notation: S is an associative K-algebra with filtration S; G is a multiplicative subgroup of S with induced filtration G; $\Lambda = \Lambda(G)$ is the graded Lie subring of $\operatorname{gr} S$ generated by the leading terms of the elements of G; and $\Gamma = \Gamma(G)$ is the "abstract" version of Λ, constructed directly in terms of G.

Since $\Lambda \subseteq \operatorname{gr} S$ and K has characteristic p, we have

$$p\Lambda = \{0\} ; \tag{5.3}$$

thus, Λ is a Lie $\mathbb{F}_p$-algebra. Equally simple, but less immediately apparent, is the fact that

$$g = 1 + u \in G^{(r)} \Rightarrow g^p = 1 + u^p \in G^{(rp)} . \tag{5.4}$$

Hence

$$\{G^{(r)}\}^p \subseteq G^{(rp)} \quad (r \geq 1) , \tag{5.5}$$

where $\{G^{(r)}\}^p$ denotes the subgroup generated by the pth powers of the elements of $G^{(r)}$. In virtue of this property, we call G a *p-filtration*. By (5.4), we have also

$$\{\Lambda_r\}^{[p]} \subseteq \Lambda_{rp} \quad (r \geq 1) , \tag{5.6}$$

where $\{\Lambda_r\}^{[p]}$ denotes the additive subgroup generated by the pth powers of the elements of Λ_r. This and Jacobson's identity show that

$$\Lambda^{[p]} \subseteq \Lambda . \tag{5.7}$$

In virtue of (5.7), Λ is called a *Lie p-subring* of $\operatorname{gr} S$; and because (5.6) holds as well, Λ is a *graded* Lie p-subring.

If H is a subgroup of G, $\Lambda(H)$ is a Lie p-subring of $\Lambda(G)$, meaning a Lie subring closed under pth powers. If N is a normal subgroup of G, $\Lambda(N)$ is a *Lie p-ideal* of $\Lambda(G)$, meaning a Lie ideal closed under pth powers; Jacobson's identity then shows that the quotient Lie ring $\Lambda(G)/\Lambda(N)$ inherits the pth power operation from $\Lambda(G)$. In order to get a clear understanding of this last result we must pass, as before, to the abstract version Γ of Λ.

We use the Lie ring isomorphism $\phi^{-1} : \Lambda \to \Gamma$ of §1 to transfer the pth power operation $u \mapsto u^p$ on Λ over to Γ. The resulting "pth power" operation $v \mapsto v^{(p)}$ on Γ is defined as follows. Let

$$v = v_1 + v_2 + \dots \in \Gamma ,$$

where

$$v_r = g_r G^{(r+1)} \in \Gamma_r \ , \quad g_r \in G^{(r)} \ .$$

Then

(5.8) $$v^{(p)} = v_1^{(p)} + v_2^{(p)} + \ldots + l(v_1, v_2, \ldots) \ ,$$

where $l(\cdot, \ldots)$ is the Lie function defined in Jacobson's identity, and

(5.9) $$\begin{aligned} v_r^{(p)} &= \left[g_r G^{(r+1)}\right]^{(p)} \\ &= g_r^p G^{(rp+1)} \in \Gamma_{rp} \ . \end{aligned}$$

The following identities, holding for all $v, w \in \Gamma$, are immediate consequences of their counterparts in Λ :

(5.10) $$(v+w)^{(p)} = v^{(p)} + w^{(p)} + l(v, w) \ ,$$

(5.11) $$[v, w^{(p)}] = [v, \underbrace{w, \ldots, w}_{p \text{ terms}}] \ .$$

In virtue of these identities, the Lie $\mathbb{F}_p$-algebra Γ , equipped with the unary operation $v \mapsto v^{(p)}$, will be called a *Lie p-ring*[9].

Suppose now that G is *any* p-filtration of G - whether obtained from a characteristic p ring filtration or not. Then we may form the graded Lie ring $\Gamma = \operatorname{gr} G$ as in §1. Moreover, (5.8) and (5.9) define unambiguously a unary operation $v \mapsto v^{(p)}$ on Γ under which it becomes a graded Lie p-ring. These results are somewhat harder to prove than the corresponding results in §1 because the group-theoretical analogues of (5.1) and (5.2) are more complicated than the group-theoretical analogues of the standard Lie ring identities.

The *lower p-central series*

(5.12) $$G = \gamma_1^p G \supseteq \gamma_2^p G \supseteq \ldots$$

is defined in terms of the lower central series by

(5.13) $$\gamma_r^p G = \prod_{\substack{s,t \\ sp^t \geq r}} (\gamma_s G)^{p^t} \ ,$$

where $(\gamma_s G)^{p^t}$ is the subgroup generated by the p^tth powers of the elements of $\gamma_s G$. It is a p-filtration with the property that

[9] Often called a *Lie p-algebra over* $\mathbb{F}_p$ or a *restricted Lie algebra over* $\mathbb{F}_p$.

(5.14) $$\gamma_r^p G \subseteq G^{(r)} \quad (r \geq 1)$$

for every p-filtration G of G. The corresponding graded Lie p-ring will be denoted by $\mathrm{gr}^p G$. It can be shown that

(5.15) *$\mathrm{gr}^p G$ is generated (as a Lie p-ring) by its first homogeneous component $\mathrm{gr}_1^p G$.*

Most of the formal results about filtrations carry over to p-filtrations with simple modifications. For example, under the general conditions of §1, we proved the Lie ring isomorphism

(1.16) $$\mathrm{gr}(G/N) \cong \Lambda(G)/\Lambda(N) .$$

When K has characteristic p, this is in fact an isomorphism of Lie p-rings: G is a p-filtration, whence G/N is a p-filtration and so the left hand side is a Lie p-ring; $\Lambda(G)/\Lambda(N)$ inherits the pth power operation of $\Lambda(G)$ (as we remarked earlier) and is in fact a Lie p-ring under the induced operation; and finally the Lie ring isomorphism preserves "pth powers".

In several respects p-filtrations behave better than general filtrations. For example, we have

LAZARD'S THEOREM[10]. *Every p-filtration of a group is induced by a filtration of a suitable associative $\mathbb{F}_p$-algebra.*

Now it is easy to show that if a p-filtration G of a group G is induced by a filtration of an associative $\mathbb{F}_p$-algebra, then

(5.16) $$D_r(\mathbb{F}_p, G) \subseteq G^{(r)} \quad (r \geq 1) ;$$

and so, by Lazard's Theorem, (5.16) holds for *every* p-filtration G of G. Comparing (5.16) with (5.14), we get the

JENNINGS-LAZARD THEOREM[11]. *For every group G,*

(5.17) $$D_r(\mathbb{F}_p, G) = \gamma_r^p G \quad (r \geq 1) .$$

This implies, in particular, that (see §4)

(5.18) $$\Delta_r(\mathbb{F}_p, G) = D_r(\mathbb{F}_p, G) \quad (r \geq 1)$$

for all groups G.

As an illustration of these results, let us consider the Magnus algebra

---

[10] Lazard [8].

[11] Proved by Jennings [4] for finite p-groups and by Lazard [8] in the general case.

$$\hat{A} = \hat{A}(\mathbb{F}_p; X) .$$

We use the notation of §§3, 4 and write

(5.19) $$P = \Lambda(F) .$$

Having $\mathbb{F}_p$ as coefficient ring brings several simplifications: subalgebras are the same as subrings; and P is mapped into itself by *every* graded endomorphism of A. Hence each homogeneous component P_r is, like A_r and L_r, an $\mathbb{F}_p\Omega$-module, where Ω, as in §3, is the monoid of all linear mappings $\theta : A_1 \to A_1$. More generally, (4.11) shows that, if N is a fully invariant subgroup of F, then $\Lambda_r(N)$ is an $\mathbb{F}_p\Omega$-submodule of P_r.

Now, by (3.12) and (5.17), $P \cong \mathrm{gr}^p F$. Hence, by (5.15),

(5.20) *P is the Lie p-subring of A generated by X.*

This enables us to identify P explicitly:

(5.21) $$P = L + L^{[p]} + L^{[p^2]} + \ldots ,$$

where $L^{[p^k]}$ is the additive group generated by the p^kth powers of the elements of L. (The right hand side of (5.21) is clearly contained in P and the identities (5.1), (5.2) show it is a Lie p-ring.)

It can be shown, using the Poincaré-Birkhoff-Witt Theorem, that, if B is a basis of L, then

(5.22) $$\tilde{B} = \{b^{p^\lambda} : b \in B, \lambda = 0, 1, \ldots\}$$

is a basis of P. From this it can be deduced that P is in fact the *free Lie p-ring on X*.

Let N be a normal subgroup of F. By (5.17)-(5.18), we have

(5.23) $$\mathrm{gr}^p(F/N) \cong P/\Lambda(N) .$$

As indicated earlier in this section, this is not merely an isomorphism of graded Lie rings but indeed an isomorphism of graded Lie p-rings. (5.20)-(5.24) are essentially due to Zassenhaus [22], who first carried over to characteristic p the programme of Magnus described in §4.

We conclude this section with a result relating the functor $\mathrm{gr}^p(\cdot)$ to varieties of groups.

PROPOSITION[12]. *Let* $\underline{\underline{V}}$ *be a variety of groups and let* $\Phi_{\underline{\underline{V}}}(X)$ *denote the free group of* $\underline{\underline{V}}$ *on the set* X . *Then the following statements are equivalent:*

(a) *the locally finite* p*-groups in* $\underline{\underline{V}}$ *form a subvariety* $\underline{\underline{W}}$;

(b) *for every positive integer* d , *there are, up to isomorphism, only finitely many finite* d*-generator* p*-groups in* $\underline{\underline{V}}$;

(c) $\mathrm{gr}^p\left(\Phi_{\underline{\underline{V}}}(X)\right)$ *is finite for every finite set* X .

Proof. *(a)* ⇒ *(b)*. If X is a set of d elements, every finite d-generator p-group in $\underline{\underline{V}}$ is a homomorphic image of the finite group $\Phi_{\underline{\underline{W}}}(X)$.

(b) ⇒ *(c)*. Let X be a finite set of d elements and write $\Psi = \Phi_{\underline{\underline{V}}}(X)$. Since each quotient group $\Psi/\gamma_r^p\Psi$ is a finite d-generator p-group in $\underline{\underline{V}}$, it follows that the sequence $\gamma_r^p\Psi$ becomes stationary. Therefore, since each quotient $\gamma_r^p\Psi/\gamma_{r+1}^p\Psi$ is finite, $\mathrm{gr}^p\Psi$ is finite.

(c) ⇒ *(a)*. For an absolutely free group Φ , set

$$\theta(\Phi) = \bigcap_r \left\{\gamma_r^p\Phi\right\}\left(v_{\underline{\underline{V}}}\Phi\right) .$$

Then θ is a subgroup functor and so, by Lemma 2 of §4, there is a variety $\underline{\underline{W}}$ such that $\theta = v_{\underline{\underline{W}}}$. Suppose now that Φ is the free group on a finite set X . The hypothesis *(c)* implies that the sequence $\left\{\gamma_r^p\Phi\right\}\left(v_{\underline{\underline{V}}}\Phi\right)$ is stationary, and $\theta(\Phi)$ is its terminal value. Since $\Phi/\theta(\Phi) \cong \Phi_{\underline{\underline{W}}}(X)$, it follows that every finitely generated group in $\underline{\underline{W}}$ is a finite p-group in $\underline{\underline{V}}$. Hence every group in $\underline{\underline{W}}$ is a locally finite p-group in $\underline{\underline{V}}$. It remains to prove the converse. Since the variety generated by a group coincides with the variety generated by its finitely generated subgroups, it is sufficient to prove that every finite p-group G in $\underline{\underline{V}}$ is in $\underline{\underline{W}}$. Represent G as a quotient Φ/N of a free group Φ , and suppose G has class c and exponent p^e . Since $G \in \underline{\underline{V}}$, $N \supseteq v_{\underline{\underline{V}}}\Phi$; and since $\gamma^p_{cp^e}G = \{1\}$, $N \supseteq \gamma^p_{cp^e}\Phi$. Therefore $N \supseteq \theta(\Phi)$ and so $G \in \underline{\underline{W}}$, as required.

6. Groups of prime exponent

The groups of exponent dividing a given prime p form a variety $\underline{\underline{V}} = \underline{\underline{B}}_p$. We are concerned here mainly with the finite groups in $\underline{\underline{V}}$. We shall work in the Magnus algebra $\hat{A} = \hat{A}\left(\mathbb{F}_p, X\right)$, using the notation of §§3-5. The free group of $\underline{\underline{V}}$ on X is

---

12 Compare Kovács [7].

isomorphic to F/F^p and we have

$$\mathrm{gr}^p(F/F^p) \cong P/\Lambda(F^p) . \tag{6.1}$$

Since F/F^p has exponent p, its lower p-central series and lower central series coincide, so that $\mathrm{gr}^p(F/F^p)$, considered as a Lie ring, is just $\mathrm{gr}(F/F^p)$.

The process known as the *linearization of identities* depends on the following simple result, which is quoted without proof.

LEMMA 1. *If a vector space over* $\mathbb{F}_p$ *is generated by elements*

$$v_{m_1,\ldots,m_s} \quad (m_i = 0, 1, \ldots, p-1; \; i = 1, \ldots, s) ,$$

then it is also generated by the elements

$$\sum_{\substack{m_1,\ldots,m_s \\ 0 \le m_i < p}} v_{m_1,\ldots,m_s} \lambda_1^{m_1} \ldots \lambda_s^{m_s} \quad \left((\lambda_1, \ldots, \lambda_s) \in (\mathbb{F}_p)^s\right) .$$

The ideal of L generated by the elements

$$[u, \underbrace{v, \ldots, v}_{r \text{ terms}}] \quad (u, v \in L) \tag{6.2}$$

is called the *rth Engel ideal* and denoted by $E(r)$.

LEMMA 2. *Define*

$$\tau_s(u_1, \ldots, u_s) = \sum_{\pi} [u_1, u_{\pi 2}, \ldots, u_{\pi s}] , \tag{6.3}$$

where summation is over the permutations π *of* $2, \ldots, s$. *If* $r < p$, $E(r)$ *is generated as a vector space by the elements*

$$\tau_{r+1}(u_0, u_1, \ldots, u_r) \quad (u_i \in L) . \tag{6.4}$$

Proof. Let U, V denote the vector spaces generated by the elements (6.2), (6.4) respectively. By inspection, the element (6.4) is the coefficient of $\lambda_1 \ldots \lambda_r$ in the expansion of

$$[u_0, \underbrace{v, \ldots, v}_{r \text{ terms}}] , \quad v = \sum_1^r \lambda_i u_i ;$$

therefore, by Lemma 1, $V \subseteq U$. On the other hand, since

$$[u, \underbrace{v, \ldots, v}_{r \text{ terms}}] = \frac{1}{r!} \tau_{r+1}(u, v, \ldots, v) ,$$

it follows that $U \subseteq V$. Hence $U = V$. Finally, the identity

$$[\tau_{r+1}(u_0, u_1, \ldots, u_r), v] = \sum_0^r \tau_{r+1}(u_0, \ldots, [u_i, v], \ldots, u_r)$$

shows that V is an ideal. Since $E(r)$ is the ideal generated by U , we have $V = E(r)$, as required.

LEMMA 3. *Define*

$$\sigma_r(u_1, \ldots, u_r) = \sum_\pi u_{\pi 1} u_{\pi 2} \ldots u_{\pi r} \tag{6.5}$$

where summation is over the permutations π *of* $1, \ldots, r$. *Then*

$$\sigma_p(u_1, \ldots, u_p) = \tau_p(u_1, \ldots, u_p) . \tag{6.6}$$

Proof. It is evident that

$$\sigma_r(u_1, \ldots, u_r) = \sum_\pi (u_1 u_{\pi 2} \ldots u_{\pi r} + u_{\pi 2} u_1 \ldots u_{\pi r} + \ldots + u_{\pi 2} \ldots u_{\pi r} u_1) ,$$

where summation is over the permutations π of $2, \ldots, r$. It is easily proved by induction on r that

$$\tau_r(u_1, \ldots, u_r) = \sum_\pi \left\{ u_1 u_{\pi 2} \ldots u_{\pi r} - \binom{r-1}{1} u_{\pi 2} u_1 \ldots u_{\pi r} + \ldots + (-1)^{r-1} \binom{r-1}{r-1} u_{\pi 2} \ldots u_{\pi r} u_1 \right\} ,$$

where summation is over the same range. The lemma now follows from the congruence

$$(-1)^i \binom{p-1}{i} \equiv 1 \pmod{p} .$$

LEMMA 4. *Let* $P^{[p]}$ *denote the subspace of* P *generated by the* p*th powers of the elements of* P . *Then* $P^{[p]}$ *is a Lie* p*-ideal of* P *and*

$$L \cap P^{[p]} = E(p-1) . \tag{6.7}$$

Proof. If B is a basis of L , then the set $\tilde{B}$ in (5.22) is a basis of P . Let $u = \sum \lambda_i b_i$, where $\lambda_i \in \mathbb{F}_p$, $b_i \in \tilde{B}$. Then

$$u^p = \sum_i (\lambda_i b_i)^p + l(\lambda_1 b_1, \ldots) ,$$

where the first part is the subspace with basis $\tilde{B} - B$ and the second in L . Hence $L \cap P^{[p]}$ is the subspace generated by the $l(\lambda_1 b_1, \ldots)$. Now, by inspection, the coefficient of $\lambda_1^{m_1} \ldots \lambda_s^{m_s}$ $(0 \le m_i < p, \sum m_i = p)$ in $l(\lambda_1 b_1, \ldots, \lambda_s b_s)$ is

$$(m_1! \ldots m_s!)^{-1} \sigma_p(\underbrace{b_1, \ldots, b_1}_{m_1 \text{ terms}}, \underbrace{b_2, \ldots, b_2}_{m_2 \text{ terms}}, \ldots) .$$

Therefore, by Lemma 1, $L \cap P^{[p]}$ is the subspace generated by the elements

$$(*) \qquad \sigma_p(c_1, \ldots, c_p) \quad (c_i \in \tilde{B}) .$$

By Lemmas 2 and 3, $E(p-1)$ is the subspace generated by those elements (*) with all c_i in B. However, by (5.2), all the remaining elements in (*) are in $E(p)$. This proves (6.7). The fact that $P^{[p]}$ is a Lie p-ideal of P follows easily from (5.2).

After these preliminaries, we are in a position to deal with $\Lambda(F^p)$. The first result is less evident than it looks:

$$(6.8) \qquad P^{[p]} \subseteq \Lambda(F^p) .$$

Proof. Let $u \in P$. Then $u = u_1 + \ldots + u_s$, where the u_i are homogeneous elements of P, and by Jacobson's identity and Lemma 4, we have $u^p \equiv u_1^p + \ldots + u_s^p$ $(\text{mod } E(p-1))$. Thus, it will be sufficient to prove the following two results:

(a) if u is a homogeneous element of P, then $u^p \in \Lambda(F^p)$;

(b) $E(p-1) \subseteq \Lambda(F^p)$.

In proving (a), we may of course assume that $u^p \neq 0$. Then $u \neq 0$ and so there exists $g = 1 + u + \ldots \in F$ with leading term u. Hence $g^p = 1 + u^p + \ldots \in F^p$ has leading term u^p, so that $u^p \in \Lambda(F^p)$ as required.

In order to prove (b), it is sufficient, by Lemmas 2 and 3, to show that $\sigma_p(v_1, \ldots, v_p) \in \Lambda(F^p)$ when $v_1, \ldots, v_p \in L$. Now by (4.10), there is no loss of generality in assuming that X contains at least p elements: let $x_1, \ldots, x_p$ be distinct elements of X. Then, by (4.12), it is sufficient to prove that $\sigma_p(x_1, \ldots, x_p) \in \Lambda(F^p)$. However, by (a),

$$(\lambda_1 x_1 + \ldots + \lambda_p x_p)^p - (\lambda_1 x_1)^p - \ldots - (\lambda_p x_p)^p \in \Lambda(F^p) .$$

Since $\sigma_p(x_1, \ldots, x_p)$ is the coefficient of $\lambda_1 \ldots \lambda_p$ in this expression, it follows from Lemma 1 that $\sigma_p(x_1, \ldots, x_p) \in \Lambda(F^p)$. This completes the proof.

Let us take stock of what has been proved up to now. Choose a basis B of L consisting of *homogeneous* elements and let $\leq$ be any total ordering of B. Write

$$(6.9) \qquad E = E(p-1) , \quad V = L \cap \Lambda(F^p) .$$

Then E is generated as a vector space by the elements $\sigma_p(b_1, \ldots, b_p)$, $b_i \in B$.

Since σ_p is a symmetric function of its p arguments and since

$\sigma_p(b, \ldots, b) = p!b^p = 0$, E is even generated by the elements

$$\sigma_p(b_1, \ldots, b_p) \text{ , } b_i \in B \text{ , } b_1 \leq \ldots \leq b_p \text{ , } b_1 < b_p \text{ .} \tag{6.10}$$

We may call these the *Kostrikin elements*[13], or briefly *K-elements*, with respect to B . It is evident that E is a graded ideal of L :

$$E = E_1 \oplus E_2 \oplus \ldots \text{ , } E_r = L_r \cap E \text{ ,} \tag{6.11}$$

and that E_r is generated as a vector space by the K-elements of weight r (that is, by the K-elements in L_r). We have of course

$$V = V_1 \oplus V_2 \oplus \ldots \text{ , } V_r = L_r \cap V \text{ ,} \tag{6.12}$$

and by (6.8),

$$E_r \subseteq V_r \quad (r \geq 1) \text{ .} \tag{6.13}$$

Finally, since, as graded Lie rings,

$$\begin{aligned} L/V &= L/L \cap \Lambda(F^p) \\ &\cong (L+\Lambda(F^p))/\Lambda(F^p) = P/\Lambda(F^p) \text{ ,} \end{aligned}$$

we have

$$\operatorname{gr}(F/F^p) \cong L/V \text{ .} \tag{6.14}$$

As mentioned in §3, each L_r is an $\mathbb{F}_p\Omega$-module, where Ω is the multiplicative monoid of all linear transformations on L_1 . It is evident that E_r is an $\mathbb{F}_p\Omega$-submodule of L_r and it follows from (4.11) that the same is true of V_r . These are useful facts, although comparatively little is known about module structure, even of L_r .

More is known about L_r and E_r as vector spaces. The *basic Lie products* of weight r form a basis of L_r , and when X is a finite set of d elements, the dimension of L_r is given by *Witt's formula*

$$\dim L_r = \frac{1}{r} \sum_{\delta | r} \mu(r/\delta) d^\delta \text{ .}$$

Similarly, an upper bound to $\dim E_r$ (when X is finite) is given by the number of K-elements of weight r . It is known that the K-elements of weight r are linearly dependent when $r > 2p - 2$ but they are conjectured to be linearly independent for

---

[13] First systematically used in Kostrikin [5].

$r \le 2p - 2$.

The deepest result in the whole subject is

KOSTRIKIN'S THEOREM[14]. *L/E is locally finite.*

By (6.14) and the proposition in §5, this implies a positive solution to the restricted Burnside problem for prime exponent:

COROLLARY 1. *For every positive integer d, there are, up to isomorphism, only finitely many finite d-generator groups of given prime exponent p.*

By the same proposition, we have:

COROLLARY 2. *The locally finite groups in $\underline{\underline{B}}_p$ form a subvariety.*

The subvariety in Corollary 2 is the *Kostrikin variety* $\underline{\underline{K}}_p$. Let $c(d, p)$ denote the nilpotency class of the d-generator free group of $\underline{\underline{K}}_p$. Then

$$c(d, 2) = 1 \quad (d \ge 2) ,$$
$$c(2, 3) = 2 , \quad c(d, 3) = 3 \quad (d \ge 3) ,$$
$$c(2, 5) = 12 ,$$

and (apart from the trivial value $c(1, p) = 1$) no other values are known. The fact that $\underline{\underline{K}}_2$ and $\underline{\underline{K}}_3$ are nilpotent varieties has no counterpart for larger primes:

RAZMYSLOV'S THEOREM[15]. *If $p \ge 5$,*

$$c(d, p) \ge 2d - 1 \quad (d \ge 1)$$

and so $\underline{\underline{K}}_p$ is not nilpotent.

Let us turn now to the problem of getting detailed information about the rth homogeneous component of $\mathrm{gr}(F/F^p)$ for relatively small values of r . Of particular interest here is the relation between E_r and V_r . It is evident that the leading term of an element of F^p has weight at least p , so that

$$V_r = E_r = \{0\} \quad (1 \le r \le p-1) .$$

More generally, we have:

SANOV'S THEOREM[16]. $V_r = E_r \quad (1 \le r \le 2p-2)$.

As a preliminary to the proof, we compare the functional equations of the exponential, and "p-truncated" exponential, functions

---

[14] Kostrikin [6].

[15] Razmyslov [15].

[16] Sanov [17].

(6.15) $$e^x = \sum_0^\infty \frac{x^r}{r!}\ ,\quad e_p(x) = \sum_0^{p-1} \frac{x^r}{r!}\ ,$$

within $\hat{A}(\mathbb{Q}, X)$. Let $x, y \in X$. By the Baker-Campbell-Hausdorff Theorem,

(6.16) $$e^x e^y = e^z\ ,$$

where $z = z(x, y) \in \hat{L}$. Let $\mathbb{Q}_{(p)}$ denote the ring of p-integers (that is, rational numbers with denominator prime to p). Then it is easy to see that

(6.17) $$e_p(x)e_p(y) = e_p(u)\ ,$$

where $u \in \hat{A}(\mathbb{Q}_{(p)}, X)$. Comparing terms of like degree in both (6.16) and (6.17), we see that

(6.18) $$u_r = z_r \quad (1 \le r \le p-1)\ ,$$

where $u_r = u_r(x, y)$ and $z_r = z_r(x, y)$ are the rth homogeneous components of u and z respectively. By (3.4), the two sides of (6.18) are in

(6.19) $$\hat{A}(\mathbb{Q}_{(p)}, X) \cap L(\mathbb{Q}, X) = L(\mathbb{Q}_{(p)}, X)\ .$$

Let us now return to $\hat{A}(\mathbb{F}_p, X)$. Since the p-truncated exponential has coefficients in $\mathbb{Q}_{(p)}$, $e_p(w)$ makes sense when $w \in \hat{\underline{\underline{a}}} = \hat{\underline{\underline{a}}}(\mathbb{F}_p, X)$. Let $\tilde{F} = \phi(F)$, where ϕ is filtered endomorphism of $\hat{A}$ such that

$$\phi(x) = e_p(x) - 1 \quad (x \in X)\ .$$

Since $\operatorname{gr}\phi$ is the identity, ϕ is an automorphism and, by (3.9),

$$\Lambda(\tilde{F}) = P\ ,\quad \Lambda(\tilde{F}^p) = \Lambda(F^p)\ .$$

It follows from (6.18), (6.19), that every element of F has the form

$$e_p(w)\ ,\quad w \in L + \hat{\underline{\underline{a}}}^{(p)}\ .$$

Now, $e_p(w)^p = e_p(w^p)$, so that $\tilde{F}^p$ is generated by elements of the form

(6.20) $$e_p(r)\ ,\quad r \in P^{[p]} + \hat{\underline{\underline{a}}}^{(2p-1)}\ .$$

It follows easily that every element of $\tilde{F}^p$ has the same form (6.20).

What we have to prove is that, if ρ is a homogeneous element of $\Lambda(\tilde{F}^p)$ of weight $s \le 2p-2$, then $\rho \in P^{[p]}$. Now, ρ is the leading term of some $e_p(r)$ in (6.20) and hence, clearly, of $1 + r$. Since $r \in P^{[p]} + \hat{\underline{\underline{a}}}^{(2p-1)}$, $\rho \in P^{[p]}$ as required.

In conclusion, we briefly indicate how the method used to prove Sanov's Theorem can be extended to give information for weight greater than $2p - 2$. The comparison of terms which yielded (6.18) gives (within $\hat{A}(\Phi, X)$)

$$u_p = z_p + \frac{1}{p!}\left((x+y)^p - x^p - y^p\right) . \tag{6.21}$$

Now, in agreement with Jacobson's identity

$$(x+y)^p - x^p - y^p = p!R(x, y) + l(x, y) , \tag{6.22}$$

where $R(x, y) \in A(\Phi_{(p)}, X)$, $l(x, y) \in L(\Phi_{(p)}, X)$. Thus, (6.21) gives

$$u_p - R(x, y) = z_p + \frac{1}{p!} l(x, y) ,$$

showing that

$$u_p - R(x, y) \in A(\Phi_{(p)}, X) \cap L(\Phi, X) = L(\Phi_{(p)}, X) .$$

Using this result, one can show that every element of the group $\tilde{F}$ in the proof of Sanov's Theorem has the form

$$e_p(w) , \quad w \in L + M + \hat{\underline{\underline{a}}}^{(p+1)} ,$$

where M is the subspace of A_r generated by the values

$$R(x_1, \ldots, x_p) \quad (x_i \in X)$$

of a certain multilinear function R. Furthermore, every element of $\tilde{F}^p$ has the form

$$e_p(r) , \quad r \in P^{[p]} + N + \hat{\underline{a}}^{(2p)} ,$$

where N is the subspace of A_{2p-1} generated by the values

$$S(x_1, \ldots, x_{2p-1}) \quad (x_i \in X)$$

of a certain multilinear function S. By actually carrying our the calculations for $p = 5, 7$, one finds that V_{2p-1} does *not* necessarily coincide with E_{2p-1} (see Wall [19]).

References

[1] N. Bourbaki, *Éléments de Mathématique*. Fasc. XXVI, *Groupes et Algèbres de Lie; Chapitre I: Algèbres de Lie* (Deuxième édition. Actualités Scientifiques et Industrielles, 1285. Hermann, Paris, 1971).

[2] N. Bourbaki, *Éléments de Mathematique*. Fasc. XXXVII, *Groupes et Algèbres de Lie*; Chapitre II: *Algèbres de Lie Libres*; Chapitre III: *Groupes de Lie* (Actualités Scientifiques et Industrielles, 1349. Hermann, Paris, 1972).

[3] Ralph H. Fox, "Free differential calculus. I. Derivation in the free group ring", *Ann. of Math.* (2) **57** (1953), 547-550.

[4] S.A. Jennings, "The structure of the group ring of a p-group over a modular field", *Trans. Amer. Math. Soc.* **50** (1941), 175-185.

[5] А.И. Кострикин [A.I. Kostrikin], "О связи между периодическими группами и кольцами Ли" [On the connection between periodic groups and Lie rings], *Izv. Akad. Nauk SSSR Ser. Mat.* **21** (1957), 289-310; *Amer. Math. Soc. Transl.* (2) **45** (1965), 165-189.

[6] А.И. Кострикин [A.I. Kostrikin], "О проблеме Бернсайда" [The Burnside problem], *Izv. Akad. Nauk SSSR Ser. Mat.* **23** (1959), 3-34; *Amer. Math. Soc. Transl.* (2) **36** (1964), 63-99.

[7] L.G. Kovács, "Varieties and the Hall-Higman paper", *Proc. Internat. Conf. Theory of Groups*, Canberra, 1965, 217-219 (Gordon and Breach, New York, London, Paris, 1967).

[8] Michel Lazard, "Sur les groupes nilpotents et les anneaux de Lie", *Ann. Sci. École Norm. Sup.* (3) **71** (1954), 101-190.

[9] Wilhelm Magnus, "Beziehungen zwischen Gruppen und Idealen in einem speziellen Ring", *Math. Ann.* **111** (1935), 259-280.

[10] Wilhelm Magnus, "Über Beziehungen zwischen höheren Kommutatoren", *J. reine angew. Math.* **177** (1937), 105-115.

[11] Wilhelm Magnus, "Über Gruppen und zugeordnete Liesche Ringe", *J. reine angew. Math.* **182** (1940), 142-149.

[12] Wilhelm Magnus, Abraham Karrass, Donald Solitar, *Combinatorial Group Theory: Presentations of Groups in Terms of Generators and Relations* (Pure and Applied Mathematics, **13**. Interscience [John Wiley & Sons], New York, London, Sydney, 1966. Second, Revised Edition: Dover, New York, 1976).

[13] Hanna Neumann, *Varieties of Groups* (Ergebnisse der Mathematik und ihrer Grenzgebiete, **37**. Springer-Verlag, Berlin, Heidelberg, New York, 1967).

[14] Peter M. Neumann, "An improved bound for BFC p-groups", *J. Austral. Math. Soc.* **11** (1970), 19-27.

[15] Ю.П. Размыслов [Ju.P. Razmyslov], "Об энгелевых алгебрах Ли" [On Engel Lie algebras], *Algebra i Logika* **10** (1971), 33-44; *Algebra and Logic* **10** (1971), 21-29.

[16] E. Rips, "On the fourth integer dimension subgroup", *Israel J. Math.* 12 (1972), 342-346.

[17] И.Н. Санов [I.N. Sanov], "Установление связи между периодическими группами с периодом простым уислом и кольцами Ли" [Establishment of a connection between periodic groups with period a prime number and Lie rings], *Izv. Akad. Nauk SSSR Ser. Mat.* 16 (1952), 23-58.

[18] M.R. Vaughan-Lee, "Breadth and commutator subgroups of p-groups", *J. Algebra* 32 (1974), 278-285.

[19] G.E. Wall, "On the Lie ring of a group of prime exponent", *Proc. Second Internat. Conf. Theory of Groups*, Canberra, 1973, 667-690 (Lecture Notes in Mathematics, 372. Springer-Verlag, Berlin, Heidelberg, New York, 1974).

[20] J. Wiegold, "Groups with boundedly finite classes of conjugate elements", *Proc. Roy. Soc. London Ser. A* 238 (1957), 389-401.

[21] Ernst Witt, "Treue Darstellung Liescher Ringe", *J. reine angew. Math.* 117 (1937), 152-160.

[22] Hans Zassenhaus, "Ein Verfahren, jeder endlichen p-Gruppe einen Lie-Ring mit der Charakteristik p zuzuordnen", *Abh. Math. Sem. Univ. Hamburg* 13 (1940), 200-207.

Department of Pure Mathematics,
University of Sydney,
Sydney,
New South Wales.

PROC. 18th SRI
CANBERRA 1978, 174-196. 20F35, 20F40

COMMUTATOR COLLECTION AND MODULE STRUCTURE

G.E. Wall

1. Introduction

Let A denote the free associative algebra over a field k on the free generators $x_1, \ldots, x_d$. We use the customary notations ab and $[a, b] = ab - ba$ for the associative and Lie products in A. Let L denote the Lie subalgebra of A generated - indeed, *freely* generated - by the x_i. As vector spaces,

$$A = \bigoplus_{r=0}^{\infty} A_r \ , \quad L = \bigoplus_{r=1}^{\infty} L_r \ ,$$

where A_r, L_r are the rth homogeneous components of A, L (the elements of A_r are said to be homogeneous of *weight* r). In particular, $A_0 = k1$ and $A_1 = L_1$ is the subspace having the x_i as basis. Let Δ denote the set of all linear transformations on A_1, regarded as a monoid under the usual composition of transformations. Since A is *freely* generated by the x_i, it follows that each element of Δ can be extended uniquely to an endomorphism of A, which evidently maps the homogeneous components of A, L into themselves. Thus, each A_r and L_r is equipped in a canonical way with a $k\Delta$-module structure.

The process of commutator collection in A yields both the sequence of basic commutators $c_1, c_2, \ldots$, forming a basis of L, and the corresponding standard monomials $c_1^{n_1}c_2^{n_2} \ldots c_s^{n_s}$ $(s \geq 0,\ n_i \geq 0)$, which form a basis of A. However, these bases are ill-adapted to the investigation of $k\Delta$-module structure, as one realizes on expressing even quite simple non-standard monomials (such as $x_d \ldots x_2x_1$) in terms of standard ones. The original aim of the present paper was to remedy this defect. This is in fact achieved when k has characteristic zero. But in characteristic $p > 0$ the methods apply essentially only to elements of weight less than p. Their usefulness in p-group theory would be greatly enhanced if the restriction to low weight could be removed.

The basic commutators of given weight are defined inductively in terms of those of smaller weight. In a similar way, we construct, in Theorems 1 and 2, direct decompositions of the vector spaces A_r and L_r, given direct decompositions

$$L_s = \bigoplus_i V_{s,i}$$

for appropriate smaller values of s. When the $V_{s,i}$ are the 1-dimensional subspaces spanned by the basic commutators of weight s (for the values of s which come into question), the resulting direct summands of A_r and L_r are the 1-dimensional subspaces spanned by the standard monomials, and basic commutators, of weight r. On the other hand, and this is the point of the constructions, if the $V_{s,i}$ are $k\Delta$-submodules, then so too are the summands of A_r and L_r.

We require some special notation. For $u_1, \ldots, u_m \in A$, we define

$$\langle u_1, \ldots, u_m \rangle = \sum_P u_{1P}u_{2P} \ldots u_{mP}, \tag{1.1}$$

summation being over all permutations P of the suffixes. It is obvious that $\langle u_1, \ldots, u_m \rangle$ is a symmetric function of its arguments. If U is a subspace of A, then $U^{\langle m \rangle}$ will denote the subspace spanned by the elements $\langle u_1, \ldots, u_m \rangle$, $u_i \in U$; $U^{\langle 0 \rangle}$ is taken to be $k1$.

We observe the convention that $\prod_{i=1}^{t} B_i$ means $B_1B_2 \ldots B_t$, even if the terms B_i do not commute. We are now in a position to state the first main result.

THEOREM 1. *Let* r *be a positive integer such that the characteristic of* k *is not a divisor of* $r!$. *Let*

$$\bigoplus_{i=1}^{r} L_i = \bigoplus_{j=1}^{s} V_j, \tag{1.2}$$

where each V_j *is a non-zero subspace of some* L_{w_j} . *Then*

$$(1.3) \qquad A_r = \oplus \prod_{j=1}^{s} V_j^{\langle m_j \rangle} ,$$

where the direct sum is taken over all choices of the integers $m_j \geq 0$ *such that* $\sum_{j=1}^{s} m_j w_j = r$. *The vector space* $\prod_1^s V_j^{\langle m_j \rangle}$ *in* (1.3) *is isomorphic to* $\bigotimes_1^s V^{(m_j)}$, *where* $V_j^{(m_j)}$ *denotes the space of symmetric tensors of rank* m_j *over* V_j . *If the* V_j *are* $k\Delta$*-modules, then the summand* $\prod_1^s V_j^{\langle m_j \rangle}$ *is a* $k\Delta$*-module isomorphic to the* $k\Delta$*-module* $\bigotimes_1^s V_j^{(m_j)}$.

Let Λ denote the algebra of all linear transformations on A . For $v \in A$, define $\mathrm{ad}\, v \in \Lambda$ by $u\, \mathrm{ad}\, v = [u, v]$ $(u \in A)$ (mappings are written on the right and composed accordingly). Since the x_i are free generators of A , there is a unique algebra homomorphism $\phi : A \to \Lambda$ such that $x_i\phi = \mathrm{ad}\, x_i$ $(1 \leq i \leq d)$. Then for $u, v \in A$, we define

$$(1.4) \qquad [u|v] = u(v\phi) .$$

The simple rules

$$(1.5) \qquad [u|1] = u , \quad [[u|v]|w] = [u|vw] ,$$

follow at once from the definition. Although by definition,

$$[u, x_i] = u(\mathrm{ad}\, x_i) = [u|x_i] ,$$

it is *not* true in general that $[u, v] = [u|v]$. However, it is easily verified that the v such that this equation holds for all u form a Lie subalgebra, whence

$$(1.6) \qquad [u|v] = [u, v] \quad \text{for } v \in L .$$

If U, V are subspaces of A , then $[U|V]$ denotes the subspace spanned by the elements $[u|v]$, $u \in U$, $v \in V$.

THEOREM 2. *Let* r *be an integer greater than or equal to* 2 *such that the characteristic of* k *is not a divisor of* $r!$. *Let*

$$(1.7) \qquad \bigoplus_{i=1}^{[\frac{1}{2}r]} L_i = \bigoplus_{j=1}^{s} V_j ,$$

where each V_j *is a non-zero subspace of some* L_{w_j} *and* $w_1 \leq w_2 \leq \ldots \leq w_s$. *Then*

(1.8) $$L_r = \bigoplus_{\alpha,\beta} \bigoplus_{m_j} \left[V_\beta \;\middle|\; \prod_{j=\alpha}^{s} V_j^{\langle m_j \rangle} \right],$$

where the outer direct sum is taken over the α, β *satisfying*

(1.9) $$1 \le \alpha \le \beta \le s,$$

(1.10) $$w_\alpha = w_\beta,$$

and the inner over all choices of the $m_j \ge 0$ *satisfying*

(1.11) $$\sum_{j=\alpha}^{t} m_j w_j > w_{t+1} - w_\alpha \quad (\alpha \le t < s),$$

(1.12) $$\sum_{j=\alpha}^{s} m_j w_j = r - w_\alpha .$$

The vector space $\left[V_\beta \;\middle|\; \prod_{\alpha}^{s} V_j^{\langle m_j \rangle} \right]$ *in* (1.8) *is isomorphic to*

$$V_\beta \otimes \left(\bigotimes_{\alpha}^{s} V_j^{(m_j)} \right)$$

if $\beta \ne \alpha$ *and to*

$$V_\alpha^{(m_\alpha, 1)} \otimes \left(\bigotimes_{\alpha+1}^{s} V_j^{(m_j)} \right)$$

if $\beta = \alpha$. *Here* $V^{(m)}$, $V^{(m,1)}$ *denote irreducible tensor spaces of types* (m) *and* $(m, 1)$ *over* V *respectively. If the* V_j *are* $k\Delta$*-modules, so are the summands in* (1.8), *and the vector space isomorphisms just cited become* $k\Delta$*-module isomorphisms.*

It is instructive to compare basic commutators $c_1, c_2, \ldots$ and standard monomials $c_{i_1} \ldots c_{i_s}$ $(i_1 \le \ldots \le i_s)$ with their $k\Delta$-module counterparts in the simplest cases. To avoid complications, let us assume that k has characteristic zero. Choosing $V_i = L_i$ $(i = 1, 2, \ldots)$ in Theorem 1, we get

$$A_1 = L_1,$$

$$A_2 = L_1^{\langle 2 \rangle} \oplus L_2,$$

$$A_3 = L_1^{\langle 3 \rangle} \oplus L_1 L_2 \oplus L_3,$$

and, in general,

(1.13) $$A_r = \oplus L_1^{\langle m_1 \rangle} \dots L_r^{\langle m_r \rangle} ,$$

where summation is over the solutions of

(1.14) $$m_1 + 2m_2 + \dots + rm_r = r$$

in non-negative integers m_i . Taking dimensions in these equations, we get the identity

$$(1-dt)^{-1} = \prod_{i=1}^{\infty} \left(1-t^i\right)^{-\left(\dim L_i\right)} \quad \left(d = \dim L_1\right) ,$$

from which the formula of E. Witt [2] for $\dim L_i$ can be derived. The more general formula of Angeline Brandt [1] for the character of L_i is obtained similarly.

A solution of (1.14) determines a partition π of r into m_1 ones, m_2 twos, and so on, and we write

(1.15) $$L^{\langle \pi \rangle} = L_1^{\langle m_1 \rangle} \dots L_r^{\langle m_r \rangle} .$$

If $c_{j_1} \dots c_{j_t}$ is a (not necessarily standard) monomial of weight r , then the weights of the individual terms c_{j_α} also form a partition of r . We shall describe a relation between $L^{\langle \pi \rangle}$ and the monomials with associated partition π .

We introduce a partial order on the set of all partitions of r by declaring that $\pi' \leq \pi$ when the parts of π are obtained by further partitioning the parts of π' . For example, if

$$\pi : 1 + 1 + 1 + 2 + 2 + 2 = 9 ,$$
$$\pi' : 1 + 3 + 5 = 9 ,$$

then $\pi' \leq \pi$ because $1 + 3 + 5 = 1 + (1+2) + (1+2+2)$. The following diagram shows what this partial order is like when $r = 5$:

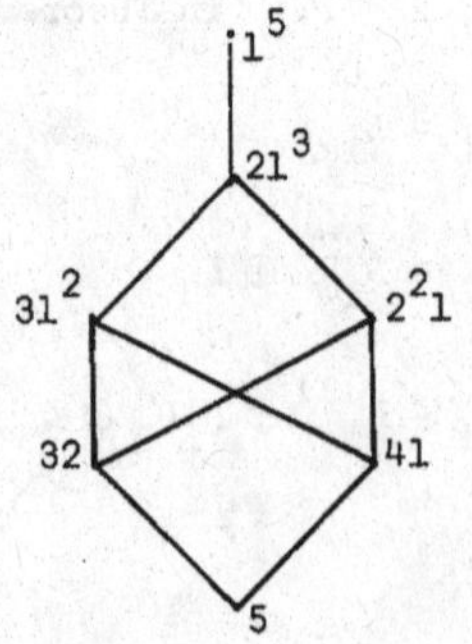

With this notation, we have the following result:

(1.16) *Let* S *be a set of partitions of* r *with the property that*

$$\pi \in S \text{ and } \pi' \le \pi \Rightarrow \pi' \in S .$$

Then every monomial $c_{j_1} \dots c_{j_t}$ *with associated partition in* S *lies in* $\bigoplus_{\pi \in S} L^{\langle \pi \rangle}$, *and the standard monomials of this kind form a basis of this space.*

The proof depends on the simple result (analogous to the lemma in §2) that if $M = c_{j_1} \dots c_{j_t}$ has associated partition π and $N = c_{k_1} \dots c_{k_t}$ is formed from M by a permutation of its terms, then

$$M \equiv N \left(\text{mod} \bigoplus_{\pi' < \pi} L^{\langle \pi' \rangle} \right) .$$

We shall not go into the details.

The situation with the $k\Delta$-module decompositions of L is more complicated. Theorem 2 tells us how to construct a $k\Delta$-module decomposition of L_r given $k\Delta$-module decompositions of $L_1, \dots, L_{[\frac{1}{2}r]}$. Starting with the decomposition $V_1 = L_1$, we may therefore construct inductively $k\Delta$-modules $V_1, V_2, \dots$ which provide decompositions of all L_r. Thus, we get

$$\begin{aligned}
L_1 &= V_1 , \\
L_2 &= V_2 = [V_1 | V_1] , \\
L_3 &= V_3 = \left[V_1 | V_1^{\langle 2 \rangle}\right] , \\
L_4 &= V_4 \oplus V_5 = \left[V_1 | V_1^{\langle 3 \rangle}\right] \oplus [V_2 | V_2] , \\
L_5 &= V_6 \oplus V_7 = \left[V_1 | V_1^{\langle 4 \rangle}\right] \oplus \left[V_1 | V_1^{\langle 2 \rangle} V_2\right] , \\
L_6 &= V_8 \oplus V_9 \oplus V_{10} \oplus V_{11} \\
&= \left[V_1 | V_1^{\langle 5 \rangle}\right] \oplus \left[V_1 | V_1^{\langle 3 \rangle} V_3\right] \oplus \left[V_2 | V_2^{\langle 2 \rangle}\right] \oplus [V_3 | V_3] ,
\end{aligned}$$

and so on. At weight 4, we could just as well have labelled $\left[V_1 | V_1^{\langle 3 \rangle}\right]$ as V_5 and $[V_2 | V_2]$ as V_4, and this choice would have led to different definitions of the V_i at later stages. Thus, there is the same element of choice in defining the sequence $V_1, V_2, \dots$ as there is in defining the sequence of basic commutators $c_1, c_2, \dots$. In a moment we shall coordinate the definitions of the two sequences so as to be able to make comparisons between them.

Just as the direct summands of A_r were parametrized by *partitions*, so the V_i may be parametrized by certain symbols which we call their *labels*. These are defined inductively as follows: the label v_1 of V_1 is the symbol a ; if $\lambda > 1$ and V_λ is the summand

$$\left[V_\beta \mid V_\alpha^{\langle m_\alpha \rangle} \dots V_s^{\langle m_s \rangle}\right]$$

in (1.8), then its label is

$$(1.17) \qquad v_\lambda = v_\beta \underbrace{v_\alpha \dots v_\alpha}_{m_\alpha \text{ terms}} \dots \underbrace{v_s \dots v_s}_{m_s \text{ terms}} ,$$

which we abbreviate to

$$v_\beta v_\alpha^{m_\alpha} \dots v_s^{m_s} ,$$

or, when $\alpha = \beta$, to

$$v_\alpha^{m_\alpha+1} \dots v_s^{m_s} .$$

Here are the labels up to weight 6 :

$$a \; ; \; a^2 \; ; \; a^3 \; ;$$

$$a^4, \; (a^2)^2 \; ; \; a^5, \; a^3a^2 \; ;$$

$$a^6, \; a^4a^2, \; (a^2)^3, \; (a^3)^2 .$$

We now define inductively the sequence of basic commutators $c_1, c_2, \dots$ in such a way that each c_j is associated with a definite one of the labels v_λ defined above. An essential ingredient in the definition is the order relation $\leq$ on the set of basic commutators of each given weight. We begin the sequence with any ordered basis of L_1 , each member of which receives the label a . Suppose now that the c_i of weight less than r , where $r > 1$, have been defined, ordered and labelled. The basic commutators of weight r are then defined in the usual way and it remains only to order and label them. For this purpose, we write a given basic commutator of weight r in the form given in Lemma 3 of §4:

$$c = [e_0 | e_1 \dots e_t] .$$

If the basic commutators $e_0, \dots, e_t$ have been labelled $\varepsilon_0, \dots, \varepsilon_t$, then c is labelled $\varepsilon_0\varepsilon_1 \dots \varepsilon_t$. The order relation $\leq$ between the basic commutators of weight r may be assigned in any way, subject only to the condition that, if c, c'

are labelled v_λ, v_μ with $\lambda < \mu$, then $c < c'$.

It can be proved by the methods of §§4, 5 that *the number of basic commutators which receive the label of* V_λ *is equal to the dimension of* V_λ . It would be interesting to have a more precise result along the lines of (1.16). The difficulty, which I have not so far been able to overcome, is to define the appropriate partial order on the set of all labels of given weight.

ACKNOWLEDGEMENT. I wish to thank L.G. Kovács and M.F. Newman for helpful discussions about this paper.

2. Proof of Theorem 1

Choose a basis $v_1, \ldots, v_{d_1}$ of V_1 , a basis $v_{d_1+1}, \ldots, v_{d_1+d_2}$ of V_2 , and so on. Then the elements

(2.1) $$v_1, v_2, \ldots, v_D \quad \left(D = \sum d_i\right)$$

form a basis of $\bigoplus_1^r L_i$ adapted to the direct decomposition (1.2). Denote the natural ordering of the elements in (2.1) by $\leq$.

Clearly, the elements $\langle u_1, \ldots, u_{m_1} \rangle$, $u_i \in \{v_1, \ldots, v_{d_1}\}$, span $V_1^{\langle m_1 \rangle}$; and since $\langle u_1, \ldots, u_{m_1} \rangle$ is a symmetric function of its arguments we may assume without loss of generality that $u_1 \leq \ldots \leq u_{m_1}$. Similar remarks apply, of course, to $V_2^{\langle m_2 \rangle}$, ... It follows that the products

(2.2) $$\langle u_1, \ldots, u_{m_1} \rangle \langle u_{m_1+1}, \ldots, u_{m_1+m_2} \rangle \ \ldots ,$$

where

(2.3) $$u_1 \leq u_2 \leq \ldots \leq u_{m_1} \leq u_{m_1+1} \leq \ldots$$

and

(2.4) $$\{u_1, \ldots, u_{m_1}\} \subseteq \{v_1, \ldots, v_{d_1}\} ,$$

$$\{u_{m_1+1}, \ldots, u_{m_1+m_2}\} \subseteq \{v_{d_1+1}, \ldots, v_{d_1+d_2}\} ,$$

$$\ldots$$

span $\prod_{1}^{s} V_j^{\langle m_j \rangle}$. Therefore the products (2.2) subject to (2.3), (2.4) and

(2.5) $$\sum_{1}^{s} m_j w_j = r$$

span the space

(2.6) $$K_r = \sum \prod_{1}^{s} V_j^{\langle m_j \rangle} ,$$

where $\sum$ is taken over the integers $m_j \geq 0$ satisfying (2.5). Since (2.5) is evidently equivalent to

(2.7) $$\sum \mathrm{wt}\; u_i = r ,$$

K_r is a subspace of A_r . We shall prove that *the elements* (2.2) *subject to* (2.3)-(2.5) *form a basis of* A_r . This implies (1.3) in Theorem 1, which is equivalent to the assertion that the sum in (2.6) is direct and $K_r = A_r$. The statement in the theorem about vector space isomorphism also follows by a simple comparison of dimensions.

Now, these elements are clearly in one-one correspondence with the products

(2.8) $$u_1 \dots u_{m_1} u_{m_1+1} \dots$$

satisfying (2.3), (2.7) and

(2.9) $$\{u_1, \dots, u_{m_1}, u_{m_1+1}, \dots\} \subseteq \{v_1, \dots, v_D\} .$$

But these products are just the standard monomials of weight r with respect to the basis (2.1) and so they form a basis of A_r . Therefore the number of elements (2.2) satisfying (2.3)-(2.5) is equal to the dimension of A_r . It will therefore be sufficient to prove that these elements span A_r , that is, that $K_r = A_r$.

LEMMA. *Let* $y_1, \dots, y_t \in L$ $(t \geq 2)$. *Then*

$$y_1 \dots y_t \equiv y_{1P} \dots y_{tP} \pmod{L^{t-1}}$$

for every permutation P *of the suffixes.*

Proof. Since every permutation is a product of adjacent transpositions, it is sufficient to prove the result for the latter. However,

$$y_1 \dots y_t - y_1 \dots y_{i-1} y_{i+1} y_i y_{i+2} \dots y_t = y_1 \dots y_{i-1} [y_i, y_{i+1}] \dots y_t \in L^{t-1}$$

and this proves the lemma.

COROLLARY. $L^t \cap A_r \subseteq L^{t-1} \cap A_r + K_r \quad (t \geq 2)$.

Proof. It is evident that $L^t \cap A_r$ is spanned by the elements $u_1 \dots u_t$, $u_i \in \{v_1, \dots, v_D\}$, $\sum \mathrm{wt}\, u_i = r$. By the lemma, $L^t \cap A_r$ is even spanned by those elements which satisfy the extra condition that $u_1 \leq \dots \leq u_t$. We take $u_1 \dots u_t$ to be the element in (2.8) and form the corresponding element, say z , in (2.2). By the lemma, and since, by the hypothesis of the theorem, $m_1!, m_2!, \dots$ are nonzero, we have

$$u_1 \dots u_t = \frac{1}{m_1!m_2!\dots} z \left(\mathrm{mod}\ L^{t-1} \cap A_r\right) .$$

Therefore $u_1 \dots u_t \in K_r + L^{t-1} \cap A_r$ and the corollary follows.

It is now easy to prove that $K_r = A_r$. Clearly, $L \cap A_r \subseteq K_r$. Hence, by the corollary, $L^t \cap A_r \subseteq K_r$ for all t . However, it is clear that $A_r = L_1^r$, whence $A_r = L^r \cap A_r \subseteq K_r$. This completes the proof of the theorem except for the statements about $k\Delta$-modules, which are dealt with in §6.

3. An Exponential Method

The simplest illustration of Theorem 1 arises from the canonical decomposition

(3.1) $$L = \bigoplus_1^\infty L_i$$

when k has characteristic 0 . We conclude (see (1.3)) that

(3.2) $$A = \bigoplus \prod_1^\infty L_i^{\langle m_i \rangle} ,$$

where the direct sum is taken over all sequences $m_1, m_2, \dots$ of non-negative integers of which only finitely many are non-zero. The problem considered here is to determine explicitly the components of a given element of A corresponding to the vector space decomposition (3.2).

For each i choose a basis

(3.3) $$v_{i1}, \dots, v_{id_i}$$

of L_i . Then

(3.4) $$v_{11}, \ldots, v_{1d_1}, v_{21}, \ldots, v_{2d_2}, \ldots$$

is a basis of L and the corresponding standard monomials

(3.5) $$\prod_{i=1}^{\infty} \prod_{j=1}^{d_i} v_{ij}^{n_{ij}}$$

form a basis of A. Assuming our element given as a linear combination of the elements (3.5), we have then to express it as a linear combination of the elements

(3.6) $$\prod_{i=1}^{\infty} \left\langle v_{i1}^{n_{i1}} \ldots v_{id_i}^{n_{id_i}} \right\rangle$$

(Of course, in both (3.5) and (3.6) the n_{ij} are non-negative integers only finitely many of which are non-zero. The notation in (3.6) is explained in (6.4) and (6.5).)

It is convenient to extend A to the algebra of formal power series in $x_1, \ldots, x_d$ over the polynomial ring

$$R = k[\ldots, \lambda_{ij}, \ldots] ,$$

where the λ_{ij} are independent indeterminates over k in one-one correspondence with the basis elements v_{ij} in (3.4). We may then form the infinite product of exponential series

(3.7) $$\prod_{i=1}^{\infty} \prod_{j=1}^{d_i} \exp\left(\lambda_{ij} v_{ij}\right) ,$$

which converges in the usual formal sense to the value

(3.8) $$\sum \left(\prod_{i=1}^{\infty} \prod_{j=1}^{d_i} \left(\lambda_{ij}^{m_{ij}} / m_{ij}! \right) \right) \left(\prod_{i=1}^{\infty} \prod_{j=1}^{d_i} v_{ij}^{m_{ij}} \right) .$$

The sum in (3.8) is over all choices of the non-negative integers m_{ij} with only finitely many non-zero.

What we have to do, in effect, is express (3.7) as an infinite product

(3.9) $$\prod_{i=1}^{\infty} \exp\left(\sum_{j=1}^{d_i} \mu_{ij} v_{ij} \right) ,$$

where the μ_{ij} are in R. For, in view of the multinomial expansion

(3.10) $$\left(\sum_1^d \mu_i v_i\right)^m = \sum_{\substack{m_i \geq 0 \\ m_1+\ldots+m_d=m}} \mu_1^{m_1} \ldots \mu_d^{m_d} \left\langle v_1^{m_1} \ldots v_d^{m_d} \right\rangle ,$$

the product (3.9) has the value

(3.11) $$\sum \left(\prod_{i=1}^{\infty} (m_i!)^{-1} \prod_{j=1}^{d_i} \mu_{ij}^{m_{ij}} \right) \left(\prod_{i=1}^{\infty} \left\langle v_{i1}^{m_{i1}} \ldots v_{id_i}^{m_{id_i}} \right\rangle \right) ,$$

where $m_i = \sum_{j=1}^{d_i} m_{ij}$ and the sum is over the same range as in (3.8). Thus, singling out the coefficient of $\prod_{i=1}^{\infty} \prod_{j=1}^{d_i} \left(\lambda_{ij}^{m_{ij}} / m_{ij}! \right)$ in (3.11), we get the expression of the element (3.5) as a linear combination of elements (3.6).

It remains to determine the μ_{ij}. Suppose the μ_{ij} have been determined for $i < s$ in such a way that

$$\left(\prod_{i=1}^{s-1} \exp\left(\sum_{j=1}^{d_i} \mu_{ij} v_{ij} \right) \right)^{-1} \left(\prod_{i=1}^{\infty} \prod_{j=1}^{d_i} \exp\left(\lambda_{ij} v_{ij} \right) \right)$$

has the form

$$\exp\left(\sum_{i=s}^{\infty} \sum_{j=1}^{d_i} \nu_{ij} v_{ij} \right) ,$$

for suitable $\nu_{ij} \in R$. Then, defining

$$\mu_{sj} = \nu_{sj} \quad (j = 1, \ldots, d_s) ,$$

we deduce from the Baker-Campbell-Hausdorff formula that

$$\left(\prod_{i=1}^{s} \exp\left(\sum_{j=1}^{d_i} \mu_{ij} v_{ij} \right) \right)^{-1} \left(\prod_{i=1}^{\infty} \prod_{j=1}^{d_i} \exp\left(\lambda_{ij} v_{ij} \right) \right)$$

has the form

$$\exp\left(\sum_{i=s+1}^{\infty} \sum_{j=1}^{d_i} \nu'_{ij} v_{ij} \right) .$$

Thus, the μ_{ij} are determined step by step. It is easily seen by induction that μ_{ij} is an isobaric polynomial of weight i in the variables λ_{ij} - in other words, it is a "homogeneous polynomial of weight i " if each of the λ_{ij} is counted as having weight i .

The process which we have described can be varied in several ways. It is not always convenient to assume that the given element is expressed initially in terms of standard monomials. This can often be circumvented by replacing the infinite product (3.7) by one more convenient. For example, if the given element were a nonstandard monomial such as $v_{11}v_{12}v_{11}$, then we could begin with $\exp(\alpha v_{11})\exp(\beta v_{12})\exp(\gamma v_{11})$, where α, β, γ are independent variables.

The method also extends to decompositions more general than (3.1) - namely, to any refinement obtained by further splitting up the L_i . It can also be extended partially to the case where k has characteristic $p > 0$. Here, in dealing with homogeneous components of weight less than p , we may use *truncated* exponential series

$$\sum_{i=0}^{p-1} \frac{x^i}{i!} .$$

4. Lemmas

The results proved here prepare the way for the proof of Theorem 2 in the next section. We use the general notation introduced in §1.

LEMMA 1. *Let* $z_1, \ldots, z_m \in L$ *, where* $m \geq 2$ *. Then*

$$\sum_{i=1}^{m} [z_i | \langle z_1, \ldots, \hat{z}_i, \ldots, z_m \rangle] = 0 . \tag{4.1}$$

Proof. The left hand side is

$$\sum_{i=1}^{m} \sum_{\substack{j=1 \\ j\neq i}}^{m} [z_i | z_j \langle z_1, \ldots, \hat{z}_i, \ldots, \hat{z}_j, \ldots, z_m \rangle] ,$$

which, in view of (1.6), is equal to

$$\sum_{\substack{i,j=1 \\ i\neq j}}^{m} [[z_i, z_j] | \langle z_1, \ldots, \hat{z}_i, \ldots, \hat{z}_j, \ldots, z_m \rangle] .$$

But this sum is zero because $[z_i, z_j] + [z_j, z_i] = 0$. This proves the lemma.

COROLLARY. *Let* V *be a subspace of* L, B *a basis of* V *and* $\leq$ *an order relation on* B *. Let* m *be an integer not less than* 2 *such that* $(m-1)!$ *is not divisible by the characteristic of* k *. Then the subspace*

$$W = [V | V^{\langle m-1 \rangle}]$$

is spanned by those elements

(4.2) $$[b_1|\langle b_2, \ldots, b_m\rangle] \quad (b_i \in B)$$

which satisfy

(4.3) $$b_1 > b_2 \le \ldots \le b_m .$$

Proof. Let K denote the set of all elements (4.2). It is evident that K spans W. Let K^* denote the set of elements (4.2) for which b_1 is *not* the least of $b_1, \ldots, b_m$. Since $\langle b_2, \ldots, b_m\rangle$ is symmetric in its arguments, the corollary asserts, in effect, that K^* spans W. Thus, it remains only to express each element (4.2) with $b_1 \le b_2 \le \ldots \le b_m$ linearly in terms of the elements of K^*.

Suppose m_1 of $b_1, \ldots, b_m$ are equal to b_1. If $m_1 = m$, the element (4.2) is obviously 0. Suppose now that $m_1 < m$; then, by the hypothesis of the corollary, m_1 is non-zero in k. On the other hand, the identity in the lemma expresses $m_1[b_1|\langle b_2, \ldots, b_m\rangle]$ as a linear combination of elements of K^*. This proves our result.

LEMMA 2. *Let* $z_1, \ldots, z_m \in L$, *where* $m \ge 2$. *Then*

(4.4) $$[z_1|\langle z_2, \ldots, z_m\rangle] \equiv (m-1)[z_1|\langle z_2, \ldots, z_{m-1}\rangle z_m] - [z_m|\langle z_2, \ldots, z_{m-1}\rangle z_1] \pmod{L''}$$

where L'' *is the second derived algebra of* L.

Proof. We use the well known (and easily proved) identity

$$\sum_{i=0}^{m-2} z^i z_m z^{m-2-i} = \sum_{j=1}^{m-1} \binom{m-1}{j} z^{m-1-j}\left[z_m|z^{j-1}\right]$$

with $z = z_2 + \ldots + z_{m-1}$. This yields

$$\left[z_1 \,\Big|\, \sum_{i=0}^{m-2} z^i z_m z^{m-2-i}\right] \equiv (m-1)\left[z_1|z^{m-2}z_m\right] - \left[z_m|z^{m-2}z_1\right] \pmod{L''},$$

since

$$\left[z_1\big|\left[z_m|z^{m-2}\right]\right] = -\left[z_m|z^{m-2}z_1\right].$$

Finally, comparing the terms which have degree 1 in each of $z_1, \ldots, z_m$ we get the lemma.

The final lemma deals with basic commutators. In effect, it gives an alternative formulation of the usual inductive definition. Let the sequence of basic commutators in L be

$$c_1, c_2, \ldots .$$

The order in which the c_i appear is the customary one used in the inductive definition; we denote it by $\leq$. We recall that each basic commutator c of weight greater than or equal to 2 has a unique standard representation $c = [c_\alpha, c_\beta]$, where c_α, c_β are basic commutators such that $c_\alpha > c_\beta$; moreover, if c_α has weight greater than or equal to 2 and its standard representation is $c_\alpha = [c_\gamma, c_\delta]$, then $c_\delta \leq c_\beta$. We call c_α and c_β the *first and second standard components* of c, respectively.

LEMMA 3. *Let* $e_0, \ldots, e_t$ *be basic commutators in* L *satisfying the following conditions:*

(4.5) $$t > 0 ,$$

(4.6) $$e_0 > e_1 \leq e_2 \leq \ldots \leq e_t ,$$

(4.7) $$\mathrm{wt}\, e_0 = \mathrm{wt}\, e_1 ,$$

(4.8) $$\mathrm{wt}[e_0|e_1 \ldots e_i] > \mathrm{wt}\, e_{i+1} \quad (0 < i < t) .$$

Then

(4.9) $$c = [e_0|e_1 \ldots e_t]$$

is a basic commutator with first and second standard components $[e_0|e_1 \ldots e_{t-1}]$ *and* e_t *respectively. Moreover, every basic commutator of weight greater than or equal to* 2 *has a unique representation in this form.*

Proof. Set $E_i = [e_0|e_1 \ldots e_i]$ $(1 \leq i \leq t)$. We prove by induction on i that E_i is basic with first and second standard components E_{i-1} and e_i respectively $(E_0 = e_0)$. The case $i = t$ will then yield the first statement in the lemma.

By definition, $E_1 = [e_0, e_1]$ and, by (4.6), $e_0 > e_1$. If $\mathrm{wt}\, e_0 \geq 2$, let $e_0 = [e_0', e_0'']$ be the standard representation. By (4.7), $\mathrm{wt}\, e_0'' < \mathrm{wt}\, e_1$ and so $e_0'' \leq e_1$. This establishes our result when $i = 1$.

Now let $i > 1$ and suppose our assertion proved for $E_{i'}$, $i' < i$. Then $E_i = [E_{i-1}, e_i]$. By induction E_{i-1} is basic with second standard component e_{i-1}. By (4.8), $\mathrm{wt}\, E_{i-1} > \mathrm{wt}\, e_i$, so that $E_{i-1} > e_i$; and by (4.6), $e_{i-1} \leq e_i$. Thus

E_i is basic with first and second standard components E_{i-1} and e_i respectively. This completes the inductive proof.

Let us show next that the representation (4.9) is unique. Suppose $c = [f_0|f_1 \dots f_u]$ is a second such representation. By what we have proved already, $[e_0|e_1 \dots e_{t-1}] = [f_0|f_1 \dots f_{u-1}]$ is the first standard component of c and $e_t = f_u$ is the second standard component of c. If $t = 1$, then by (4.7), $\mathrm{wt}\, e_0 = \mathrm{wt}\, e_1$; hence $\mathrm{wt}[f_0|f_1 \dots f_{u-1}] = \mathrm{wt}\, f_u$ and so, by (4.8), $u = 1$; thus $e_0 = f_0$. If $t > 1$, then the argument just given shows that also $u > 1$; in this case, we may assume by induction on the weight that $u - 1 = t - 1$ and $f_i = e_i$ $(0 \le i \le t-1)$. This proves uniqueness.

Finally, we have to show that each basic commutator c of weight greater than or equal to 2 has a representation (4.9). The proof is again by induction on the weight. Let $c = [c_\alpha, c_\beta]$ be the standard representation. If $\mathrm{wt}\, c_\alpha = \mathrm{wt}\, c_\beta$, then we get a representation (4.9) by taking $t = 1$, $e_0 = c_\alpha$, $e_1 = c_\beta$. On the other hand, if $\mathrm{wt}\, c_\alpha > \mathrm{wt}\, c_\beta$, then *a fortiori* $\mathrm{wt}\, c_\alpha \ge 2$ and so, by induction, $c_\alpha = [e_0|e_1 \dots e_{t-1}]$, where the basic commutators $e_0, \dots, e_{t-1}$ obey those parts of (4.5)-(4.8) which refer to them. Taking e_t as c_β, we see that c has a representation (4.9), as required. This completes the proof of Lemma 3.

5. Proof of Theorem 2

This follows the same general lines as the proof of Theorem 1 but is rather more elaborate in detail.

Choose bases $v_1, \dots, v_{d_1}$ of V_1, $v_{d_1+1}, \dots, v_{d_1+d_2}$ of V_2, and so on, and put these together to form the basis

$$(5.1) \qquad v_1, \dots, v_D \quad \left(D = \sum_1^s d_i\right)$$

of $\bigoplus_1^{[\frac{1}{2}r]} L_i$. Denote the natural ordering of the elements in (5.1) by $\le$. By the hypotheses of the theorem, the elements (5.1) are homogeneous and appear in order of increasing weight. Clearly, the first D basic commutators, say

$$(5.2) \qquad c_1, \dots, c_D ,$$

also form a basis of $\bigoplus_1^{[\frac{1}{2}r]} L_i$. Clearly, we have

(5.3) $$\mathrm{wt}\ v_i = \mathrm{wt}\ c_i \quad (1 \le i \le D) .$$

Consider now a typical summand

(5.4) $$\left[V_\beta \,\middle|\, \prod_{j=\alpha}^{s} V_j^{\langle m_j \rangle} \right]$$

in (1.8). It is not necessary yet to assume that all the conditions (1.9)-(1.13) hold but merely that (1.9) holds and

(5.5) $$m_\alpha > 0 .$$

That (5.5) is a consequence of the other conditions is seen as follows: if $\alpha < s$, (1.11) with $t = \alpha$ gives $m_\alpha w_\alpha > w_{\alpha+1} - w_\alpha \ge 0$; if $\alpha = s$, (1.12) gives $m_\alpha w_\alpha = r - w_\alpha \ge r - [\tfrac{1}{2}r] > 0$.

It is evident that the subspace (5.4) is spanned by the elements

(5.6) $$[u_0 | \langle u_1, \ldots, u_{m_\alpha} \rangle \langle u_{m_\alpha+1}, \ldots, u_{m_\alpha+m_{\alpha+1}} \rangle \ldots] ,$$

where the u_i are basis elements (5.1) satisfying the conditions

(5.7) $$\begin{cases} u_0 \in V_\beta , \\ u_1, \ldots, u_{m_\alpha} \in V_\alpha , \\ u_{m_\alpha+1}, \ldots, u_{m_\alpha+m_{\alpha+1}} \in V_{\alpha+1} , \\ \ldots \end{cases}$$

and

(5.8) $$\begin{cases} u_1 \le \ldots \le u_{m_\alpha} , \\ u_{m_\alpha+1} \le \ldots \le u_{m_\alpha+m_{\alpha+1}} , \\ \ldots . \end{cases}$$

Now we may in fact impose the extra restriction that

(5.9) $$u_0 > u_1 .$$

For, by (5.5), $u_1 \in V_\alpha$. Hence, if $\beta > \alpha$, (5.9) is automatically satisfied. On the other hand, if $\beta = \alpha$, (5.9) is no longer automatically satisfied but may be assumed to hold by the corollary to Lemma 1.

Since $u_{m_\alpha} < u_{m_\alpha+1}$, and so on, we may condense (5.8) and (5.9) into the one line of inequalities

(5.10) $$u_0 > u_1 \le u_2 \le \ldots \le u_{m_\alpha} \le u_{m_\alpha+1} \le \ldots \le u_M \quad \left(M = \sum_{j=\alpha}^{s} m_j\right) .$$

We remark now that (1.10)-(1.12) are equivalent to the following conditions:

(5.11) $$\mathrm{wt}\, u_0 = \mathrm{wt}\, u_1 ,$$

(5.12) $$\mathrm{wt}\left[u_0 | u_1 \ldots u_i\right] > \mathrm{wt}\, u_{i+1} \quad (1 \le i < M) ,$$

(5.13) $$\mathrm{wt}\left[u_0 | u_1 \ldots u_M\right] = r .$$

Here (5.11) and (5.13) are evidently direct transcriptions of (1.10) and (1.12) respectively. In transcribing (1.11), we note that the "t" case of (1.11) is a consequence of the "$t+1$" case when $m_{t+1} = 0$ and $t + 1 = s$. We also note that, when $m_s = 0$, the "$s-1$" case of (1.11) is a consequence of (1.12):

$$\sum_{\alpha}^{s-1} m_j w_j = \sum_{\alpha}^{s} m_j w_j = r - w_\alpha > w_s - w_\alpha ,$$

since $w_s \le [\frac{1}{2}r]$. The upshot of these remarks is that the inequality in (1.11) has only to be imposed for those values of t such that $m_{t+1} > 0$. But these last conditions mean that the inequality in (5.12) holds for those values of i of the form $m_\alpha + m_{\alpha+1} + \ldots + m_t$ with $\alpha \le t < s$ and $m_{t+1} > 0$; in short, for those i such that $\mathrm{wt}\, u_i < \mathrm{wt}\, u_{i+1}$. Since the inequality in (5.12) is obvious when $\mathrm{wt}\, u_i = \mathrm{wt}\, u_{i+1}$, it follows that (1.10)-(1.12) are equivalent to (5.11)-(5.13), as asserted.

Let

(5.14) $$T_r = \sum_{\alpha,\beta} \sum_{m_j} \left[V_\beta \,\middle|\, \prod_{j=\alpha}^{s} V_j^{\langle m_j \rangle} \right] ,$$

where summation is over the same range as in (1.8). Let B_r denote the set of all elements (5.6) in which the u_i are basis elements (5.1) satisfying (5.7) and (5.10)-(5.13). Clearly, B_r spans T_r and if B_r is a basis of T_r then the double sum in (5.14) is direct. (In saying that B_r is a basis, we mean to imply also that *formally* distinct elements of B_r - that is, ones corresponding to different choices of the u_i - are *actually* distinct as elements of A.) We shall prove that *B_r is a basis of L_r*. This implies the decomposition (1.8) in the theorem. It also gives, on comparing dimensions, the statements about vector space isomorphisms (this is amplified in §6).

In the natural one-one correspondence $v_i \leftrightarrow c_i$ $(i = 1, \ldots, D)$ between the basis elements (5.1) and basic commutators (5.2), let the elements $u_0, \ldots, u_M$ for a particular element (5.6) correspond to $e_0, \ldots, e_M$ respectively and form the element

(5.15) $$[e_0|e_1 \ldots e_M] .$$

It follows from Lemma 3 that, as the element (5.6) runs over B_r , the corresponding element runs precisely over the basic commutators of weight r . Hence, the number of formally distinct elements in B_r is $\dim L_r$. Thus, in order to prove that B_r is a basis of L_r it is sufficient to prove that B_r *spans* L_r , that is, that $T_r = L_r$.

It is clearly sufficient to prove that each Lie product $[v, u]$, where $v \in L_{r-w}$, $u \in L_w$, $r \geq 2w$, is in T_r . Since $w \leq [\frac{1}{2}r]$, we may assume that u is one of the basis elements (5.1); let $u \in V_\alpha$. We may assume, by induction, that a Lie product $[v', u']$ of the same kind is in T_r when $u' \in V_{\alpha'}$ with $\alpha' > \alpha$. We may also assume, by a second induction, that *either*

(i) v is one of the basis elements (5.1) (this can only occur when $w = \frac{1}{2}r$) *or*

(ii) v has the form

$$[u_0|\langle u_1, \ldots\rangle \ldots \langle \ldots, u_M\rangle] \quad (M \geq 1)$$

in (5.6), where the u_i are basis elements (5.1) satisfying (5.10)-(5.12) and, of course, $\operatorname{wt} v = r - w$.

We write this in the more convenient form

$$v = [z_1|\langle z_2, \ldots, z_{m-1}\rangle] \quad (m \geq 2) ,$$

where $z_2, \ldots, z_{m-1}$ are basis elements (5.1) lying in the one subspace V_λ and *either*

(a) $\operatorname{wt} z_1 > \operatorname{wt} z_2$ *or*

(b) z_1 is also a basis element (5.1) which has the same weight as the other z_j and lies in V_μ with $\mu \geq \lambda$.

Case (i) is easily disposed of. If, as we may assume, $v > u$, then $[v, u]$ is, by its form, in B_r and so in T_r.

Consider now case (ii). If $\alpha > \lambda$, then

$$[v, u] = [z_1|\langle z_2, \ldots, z_{m-1}\rangle u]$$

is, by its form, in B_r and so in T_r (notice that

$$\mathrm{wt}[z_1|\langle z_2, \ldots, z_{m-1}\rangle] = r - w > \mathrm{wt}\, u = w\ ).$$

If $\alpha < \lambda$, we write $[v, u]$ in the form

$$\sum_{i=2}^{m-1} [t_i, z_i, u] = -\sum_{i=2}^{m-1} [[z_i, u], t_i] - \sum_{i=2}^{m-1} [[u, t_i], z_i] ,$$

where

$$t_i = [z_1|\langle z_2, \ldots, \hat{z}_i, \ldots, z_{m-1}\rangle] .$$

Since each of $[z_i, u]$, $[u, t_i]$, t_i, z_i either has weight greater than w or is a basis element (5.1) in a V_ω with $\omega > \alpha$, this sum is in T_r by our induction hypothesis.

Finally, let $\alpha = \lambda$. We note that, by the induction hypothesis, $\Lambda'' \cap L_r \subseteq T_r$, where Λ is the Lie algebra generated by $z_1, \ldots, z_{m-1}, u$. Therefore, by Lemma 2,

$$\begin{aligned}(m-1)[v, u] &\equiv [z_1|\langle z_2, \ldots, z_{m-1}, u\rangle] + [u|\langle z_2, \ldots, z_{m-1}\rangle z_1] \\ &\equiv [u|\langle z_2, \ldots, z_{m-1}\rangle z_1] \pmod{T_r} ,\end{aligned}$$

since $[z_1|\langle z_2, \ldots, z_{m-1}, u\rangle]$ is, by its form, in B_r. We have $\mathrm{wt}[u|\langle z_2, \ldots, z_{m-1}\rangle] > w$ as $m \geq 2$; hence, if case (a) applies or case (b) applies and $\mu > \lambda$, then $[u|\langle z_2, \ldots, z_{m-1}\rangle z_1] \in T_r$ and so $(m-1)[v, u] \in T_r$. As $m - 1 < r$, $m - 1$ is non-zero by the hypothesis of the theorem and so $[v, u] \in T_r$.

There remains the case where $z_1 \in V_\lambda$. Interchanging the roles of u and z_1, we have

$$(m-1)[u|\langle z_2, \ldots, z_{m-1}\rangle z_1] \equiv [v, u] \pmod{T_r} .$$

Hence

$$(m-1)^2[v, u] \equiv [v, u] \pmod{T_r} ;$$

that is

$$m(m-2)[v, u] \equiv 0 \pmod{T_r}.$$

Since, by the hypothesis of the theorem, $r! \neq 0$ in k and $m \leq r$, we conclude that $[v, u] \in T_r$. This completes the proof except for the statements about $k\Delta$-modules, which are dealt with in §6.

6. Module Structure

It was pointed out in §1 that, if the *given* summands V_i in Theorems 1 and 2 are $k\Delta$-modules, then so are the *resulting* summands in (1.3) and (1.8); the same is true, a little more generally, if Δ is replaced by any submonoid Γ (for example, the general and special linear groups on L_1). The point which we wish to make here is that the *proofs* (as opposed to the *statements*) of the two theorems give specific information about the nature of the summands in (1.3) and (1.8).

In the first place, it is clearly sufficient to examine the individual summands in (1.3) and (1.8). By the nature of the bases obtained in the proofs,

$$\prod_{j=1}^{s} V_j^{\langle m_j \rangle} \cong \bigotimes_{j=1}^{s} V_j^{\langle m_j \rangle}, \tag{6.1}$$

$$\left[V_\beta \,\middle|\, \prod_{j=\alpha}^{s} V_j^{\langle m_j \rangle}\right] \cong V_\beta \otimes \left(\bigotimes_{j=\alpha}^{s} V_j^{\langle m_j \rangle}\right) \tag{6.2}$$

when $\beta > \alpha$ and

$$\left[V_\alpha \,\middle|\, \prod_{j=\alpha}^{s} V_j^{\langle m_j \rangle}\right] \cong \left[V_\alpha \,\middle|\, V_\alpha^{\langle m_\alpha \rangle}\right] \otimes \left(\bigotimes_{j=\alpha}^{s} V_j^{\langle m_j \rangle}\right). \tag{6.3}$$

We recall that the numerical constraints (1.9)-(1.12) imposed in Theorem 2 imply that $m_\alpha > 0$ in (6.2) and (6.3) (see (5.5)). The isomorphisms in (6.1)-(6.3) are $k\Gamma$-module isomorphisms, the factors $V_j^{\langle m_j \rangle}$ and $\left[V_\alpha \,\middle|\, V_\alpha^{\langle m_\alpha \rangle}\right]$ evidently being $k\Gamma$-modules. It remains to elucidate the structure of the latter.

Let us consider, in general, a subspace V of some homogeneous component L_w. We assume V is a $k\Gamma$-submodule of L_w. Now let m be a positive integer. We define $V^{\langle m \rangle}$, a little more generally than before, as the subspace spanned by all the elements

$$\left\langle a_1^{m_1} \ldots a_r^{m_r} \right\rangle, \quad \sum m_i = m, \quad a_i \in V, \tag{6.4}$$

where the element (6.4) denotes the sum of all formally distinct products $q_1 q_2 \cdots q_m$ in which m_1 of the q_i are equal to a_1, m_2 of the q_i are equal to a_2, and so on. In the case where $m!$ is not divisible by the characteristic of k, this definition of $V^{\langle m\rangle}$ coincides with the old one, for then

$$(6.5)\qquad \left\langle a_1^{m_1} \ldots a_r^{m_r}\right\rangle = \frac{1}{m_1!\ldots m_r!}\langle \underbrace{a_1, \ldots, a_1}_{m_1 \text{ terms}}, \ldots, \underbrace{a_r, \ldots, a_r}_{m_r \text{ terms}}\rangle .$$

Now we may form the *abstract version* $V^{(m)}$ of $V^{\langle m\rangle}$ as follows: we take the tensor algebra A^* of V, which is the same thing as the free associative algebra on a basis of V, and we take $V^{(m)}$ as the subspace of A^* spanned by the elements (6.4) *calculated in* A^*; in short, $V^{(m)}$ is the space of symmetric tensors of rank m on the vector space V. Now $V^{(m)}$ inherits the $k\Gamma$-module structure from V and there is clearly a $k\Gamma$-module homomorphism $\theta : V^{(m)} \to V^{\langle m\rangle}$. What the proof of Theorem 1 implies is that, when $m!$ *is not divisible by the characteristic of* k, θ *is an isomorphism.* Indeed, the proof shows that the elements

$$\langle u_1, \ldots, u_m\rangle , \quad u_1 \le \ldots \le u_m ,$$

where the u_i run over an ordered basis of V, form a basis of $V^{(m)}$ or $V^{\langle m\rangle}$, according to which multiplication is used.

Similar considerations apply to $[V|V^{\langle m\rangle}]$. We take V, as before, to be a $k\Gamma$-submodule of some L_w and we assume that $(m+1)!$ is not a multiple of the characteristic of k. The symmetric group on the symbols $0, 1, \ldots, m$ operates on the $(m+1)$th power of V in A^* by

$$(v_0 v_1 \ldots v_m)P = v_{0P^{-1}} \ldots v_{mP^{-1}} .$$

We introduce the symmetry operator

$$\Omega = \Delta\big(I-(0,\ 1)\big) ,$$

where Δ is the sum of all permutations which fix 0, I is the identity and $(0,\ 1)$ the transposition interchanging 0 and 1. Apart from the scalar factor $2!m!$, Ω is the primitive idempotent corresponding to the Young tableau

$$\begin{pmatrix} 1 & 2 & \ldots & m \\ 0 & & & \end{pmatrix} .$$

Set

$$\phi(v_0, \ldots, v_m) = (v_0 \ldots v_m)\Omega .$$

Since

$$\begin{aligned}(v_0 \ldots v_m)\Delta &= v_0\langle v_1, \ldots, v_m\rangle \\ &= \sum_{i=1}^{m} v_0 v_i \langle v_1, \ldots, \hat{v}_i, \ldots, v_m\rangle ,\end{aligned}$$

we have

$$\phi(v_0, \ldots, v_m) = \sum_{i=1}^{m} [v_0, v_i]\langle v_1, \ldots, \hat{v}_i, \ldots, v_m\rangle .$$

Since

$$P\Omega = \Omega$$

if the permutation P fixes 0 , and

$$\big(I+(0, 1) + \ldots + (0, m)\big)\Omega = 0 ,$$

it follows that $\phi(v_0, \ldots, v_m)$ is symmetric in $v_1, \ldots, v_m$ and satisfies

$$\sum_{i=0}^{m} \phi(v_i, v_0, v_1, \ldots, \hat{v}_i, \ldots, v_m) = 0 .$$

Set

$$V^{(m,1)} = \{\phi(v_0, \ldots, v_m) : v_i \in V\} .$$

Then both $V^{(m,1)}$ and $[V|V^{\langle m\rangle}]$ are $k\Gamma$-modules and

$$\phi(v_0, \ldots, v_m) \mapsto [v_0|\langle v_1, \ldots, v_m\rangle]$$

defines a $k\Gamma$-module isomorphism. We omit the straightforward details.

References

[1] Angeline Brandt, "The free Lie ring and Lie representations of the full linear group", *Trans. Amer. Math. Soc.* 56 (1944), 528-536.

[2] E. Witt, "Treue Darstellung Lieschen Ringe", *J. Reine Angew. Math.* 177 (1937), 152-160.

Department of Pure Mathematics,
University of Sydney,
Sydney,
New South Wales.

PROC. 18th SRI
CANBERRA 1978, 197-204.

INDUCED REPRESENTATIONS OF LIE ALGEBRAS

William H. Wilson

Introduction

This article describes axiomatic approaches to induced representations of Lie algebras, and includes references to similar work on permutation groups and coalgebras.

Known families of finite-dimensional simple modules for semi-simple complex Lie algebras are constructed by an induction-like process (Humphreys [2], p. 109), yet the standard induction functors always give rise to infinite-dimensional induced modules.

In §1, suitable properties for induced modules are described. Proposition 1 shows that certain adjoint-like induced modules are always infinite-dimensional. In §2, weakened adjunctions are introduced, and it is shown that these may correspond to suitable induction functors; examples are outlined. In §3, another type of adjoint-like induced module functor is considered. Theorem 2 relates the existence of such functors to conditions on the finite-dimensional sub- and quotient- modules of the standard induced modules.

1. Properties of induced modules

In the theory of representations of finite groups on vector spaces, there are two natural candidates for the notion of an induced module. Specifically, if $H < G$ are finite groups, k is a field, and M is a finite-dimensional right kH-module, then $M \otimes_{kH} kG$ and $\hom_{kH}(kG, M)$ are isomorphic, finite-dimensional kG-modules "induced" by M .

Let us contrast this with the situation for finite-dimensional Lie algebras. Let $\underline{\underline{h}} < \underline{\underline{g}}$ be Lie algebras of finite dimension over a field k , let $U\underline{\underline{h}}$ and $U\underline{\underline{g}}$ be the universal enveloping algebras of $\underline{\underline{h}}$ and $\underline{\underline{g}}$, and let W be a non-zero finite-dimensional right $\underline{\underline{h}}$-module (or, equivalently, $U\underline{\underline{h}}$-module). Then $W \otimes_{U\underline{\underline{h}}} U\underline{\underline{g}}$ and $\hom_{U\underline{\underline{h}}}(U\underline{\underline{g}}, W)$ are non-isomorphic $U\underline{\underline{g}}$-modules "induced" by W , and both are infinite-dimensional, since $U\underline{\underline{g}}$ is a free $U\underline{\underline{h}}$-module of countable rank (see Humphreys [2], p. 92). On the other hand, there are induced-module constructions, for particular pairs of Lie algebras $\underline{\underline{h}}$ and $\underline{\underline{g}}$, which produce finite-dimensional induced $\underline{\underline{g}}$-modules for certain $\underline{\underline{h}}$-modules, as noted in the introduction.

This leads us to ask if there are other ways of constructing, from W , an induced $\underline{\underline{g}}$-module with, perhaps, more tractable properties. Now we are faced with a question: what are suitable properties for a $\underline{\underline{g}}$-module induced from W ? The following is a list of properties that we might consider desirable:

(i) the induced module should depend functorially on W ;

(ii) the induced module should contain W as an $\underline{\underline{h}}$-submodule;

(iii) the inclusion of W in the induced module should be natural;

(iv) W should generate the induced module as $U\underline{\underline{g}}$-module;

(v) if W is finite-dimensional, then the induced module should be finite-dimensional, at least sometimes;

(vi) analogues of the Frobenius reciprocity laws should hold.

We could also require that the duals of properties (ii), (iii) and (iv) hold, that is, that:

(ii)' the induced module should have W as an $\underline{\underline{h}}$-quotient-module;

(iii)' the projection from the induced module to W should be natural;

(iv)' the kernel of the projection should contain no non-zero $\underline{\underline{g}}$-modules.

We shall denote by mod-$\underline{\underline{h}}$ and mod-$\underline{\underline{g}}$ the categories of all right $\underline{\underline{h}}$- (or $U\underline{\underline{h}}$-) and $\underline{\underline{g}}$- (or $U\underline{\underline{g}}$-) modules. F : mod-$\underline{\underline{g}} \to$ mod-$\underline{\underline{h}}$ will denote the forgetful functor. IW will denote the $\underline{\underline{g}}$-module induced by $W \in$ mod-$\underline{\underline{h}}$. Thus, when assumption (i) is in force, I : mod-$\underline{\underline{h}} \to$ mod-$\underline{\underline{g}}$ is an induction functor.

Property (vi) is motivated by the fact that it holds for the "usual" induced modules $W \otimes_{U\underline{\underline{h}}} U\underline{\underline{g}}$ and $\hom_{U\underline{\underline{h}}}(U\underline{\underline{g}}, W)$ in the sense that, for all $V \in$ mod-$\underline{\underline{g}}$ and $W \in$ mod-$\underline{\underline{h}}$, there are natural bijections

(vi) (l) $\hom_{U\underline{\underline{h}}}(W, FV) \to \hom_{U\underline{\underline{g}}}\left(W \otimes_{U\underline{\underline{h}}} U\underline{\underline{g}}, V\right)$, and

(vi) (r) $\hom_{U\underline{\underline{h}}}(FV, W) \to \hom_{U\underline{\underline{g}}}\left(V, \hom_{U\underline{\underline{h}}}(U\underline{\underline{g}}, W)\right)$.

Unfortunately, property (vi) (l) or (r) uniquely determines the induced module IW in the following sense:

PROPOSITION 1 (compare Mac Lane [4], p. 232). *If, for all* $V \in$ mod-$\underline{g}$, *there is a natural bijection*

$$\eta_V : \hom_{U\underline{g}}(V, IW) \to \hom_{U\underline{h}}(FV, W) ,$$

then $IW \simeq \hom_{U\underline{h}}(U\underline{g}, W)$ *as* $U\underline{g}$*-modules. Dually, if for all* $V \in$ mod-$\underline{g}$ *there is a natural bijection*

$$\varepsilon_V : \hom_{U\underline{g}}(IW, V) \to \hom_{U\underline{h}}(W, FV)$$

then $IW \simeq W \otimes_{U\underline{h}} U\underline{g}$ *as* $U\underline{g}$*-modules.*

Proof. Compose η_V with the natural bijection (vi) (r), and call the resulting bijection

$$\bar{\eta}_V : \hom_{U\underline{g}}(V, IW) \to \hom_{U\underline{g}}\big(V, \hom_{U\underline{h}}(U\underline{g}, W)\big) .$$

Let us write RW as an abbreviation for $\hom_{U\underline{h}}(U\underline{g}, W)$. Set $\lambda_W = \bar{\eta}_{IW}(1_{IW}) : IW \to RW$ and $\rho_W = \bar{\eta}^{-1}_{RW}(1_{RW}) : RW \to IW$. By the naturality of $\bar{\eta}_V$, $\bar{\eta}_V(\phi) = \lambda_W \circ \phi$ for $\phi \in \hom_{U\underline{g}}(V, IW)$, and $\bar{\eta}^{-1}_V(\chi) = \rho_W \circ \chi$ for $\chi \in \hom_{U\underline{g}}(V, RW)$. Thus $1_{IW} = \bar{\eta}^{-1}_{IW}(\bar{\eta}_{IW}(1_{IW})) = \rho_W \circ \lambda_W$ and $1_{RW} = \bar{\eta}_{RW}\big(\bar{\eta}^{-1}_{RW}(1_{RW})\big) = \lambda_W \circ \rho_W$. So $\lambda_W : IW \to RW$ is a $U\underline{g}$-isomorphism.

The proof of the dual statement is similar. □

2. Weakened adjunctions in representation theory

Despite Proposition 1, it may be possible to retain property (vi) of §1 in a weakened form. There are two possibilities at least: we can modify the categories involved (this strategy is examined in §3), or we can weaken the requirement that the map η_V or ε_V in Proposition 1 should be bijective (but in this context we shall require that η_V and/or ε_V be natural in W as well as in V).

DEFINITION. Suppose that I : mod-$\underline{h} \to$ mod-$\underline{g}$ is a functor and that there exist injections

(1) $$\eta_{VW} : \hom_{U\underline{g}}(V, IW) \to \hom_{U\underline{h}}(FV, W)$$

(respectively

(2) $$\varepsilon_{WV} : \hom_{U\underline{g}}(IW, V) \to \hom_{U\underline{h}}(W, FV) \;)$$

natural in V and W . Then we say that I is an *injective weak right* (respectively *left) adjoint* to F .

Clearly the functors I and F in this definition could be between any two categories: $\underline{\underline{H}} \underset{F}{\overset{I}{\rightleftarrows}} \underline{\underline{G}}$. The obvious surjective dual concept of weakened adjunction has been studied by Kainen [3]. The theories in the two cases turn out to be quite different: in effect, Kainen's surjective weak adjoints satisfy existence conditions whereas the injective weak adjoints satisfy uniqueness conditions.

Given a natural injection (1) or (2), it is natural to put $V = IW$ and consider the image of 1_{IW} , much as in the proof of Proposition 1. Let us write

$$(3) \qquad \begin{cases} j_W = \varepsilon_{W,IW}(1_{IW}) \in \hom_{U\underline{\underline{h}}}(W, FIW) , \\ \text{and} \\ d_W = \eta_{IW,W}(1_{IW}) \in \hom_{U\underline{\underline{h}}}(FIW, W) . \end{cases}$$

Now suppose that I is a faithful functor (that is, that I is injective on morphisms). It can be shown (see Wilson [9]) that there exist natural injections ε_{WV} as in (2) if and only if conditions (ii), (iii), and (iv) of §1 are satisfied with j_W as the natural inclusion, and dually for η_{VW} (as in (1)), conditions (ii)', (iii)', and (iv)' of §1, and d_W .

EXAMPLE. The concepts described above were suggested partly by the axiomatic representation theory of Green [1], and partly by properties of a functor described by Wallach [7, 8]. Wallach's induced module W^* was constructed by embedding a copy of W (an arbitrary $\underline{\underline{h}}$-module, but with a special class of pairs $\underline{\underline{h}}$ and $\underline{\underline{g}}$ of Lie algebras) in $F\left(\hom_{U\underline{\underline{h}}}(U\underline{\underline{g}}, W)\right)$ in a certain way, and then defining W^* to be the $U\underline{\underline{g}}$-submodule of $\hom_{U\underline{\underline{h}}}(U\underline{\underline{g}}, W)$ generated by the embedded copy of W . The functor $W \to W^*$ satisfies conditions (1) and (2), and furthermore $d_W \circ j_W = 1_W$ in the notation of (3). Wallach proves that if $\underline{\underline{g}}$ is a semi-simple complex Lie algebra and $\underline{\underline{h}}$ is a Cartan subalgebra of $\underline{\underline{g}}$, and W is a 1-dimensional $\underline{\underline{h}}$-module determined by a dominant integral weight, then W^* is a finite-dimensional simple $\underline{\underline{g}}$-module. (In fact, he proves much more than this.)

A dual construction is possible: one defines a certain $\underline{\underline{h}}$-epimorphism $\hat{d}_W : F\left(W \otimes_{U\underline{\underline{h}}} U\underline{\underline{g}}\right) \to W$ and then defines IW to be the quotient module of $W \otimes_{U\underline{\underline{h}}} U\underline{\underline{g}}$ formed by factoring out the unique largest $U\underline{\underline{g}}$-submodule of $\ker \hat{d}_W$. This construction is functorial, the functor I satisfies (1) and (2), and $d_W \circ j_W = 1_W$ in the notation of (3). Again, the induced module is finite-dimensional in the classical situation described above. (See Humphreys [2], §20, for what is essentially a proof of this.)

There are other examples of injective weak adjoints in representation theory. For permutation representations of finite groups, see Wilson [9], and for corepresentations of coalgebras, see Trushin [6] and Wilson [9]. For representations of Lie algebras possessing complemented ideals, see Wilson [10].

3. The categories of finite-dimensional $\underline{\underline{h}}$- and $\underline{\underline{g}}$- modules

Recall that mod-$\underline{\underline{g}}$ and mod-$\underline{\underline{h}}$ denote the categories of all (possibly infinite-dimensional) right modules for finite-dimensional Lie algebras $\underline{\underline{h}} < \underline{\underline{g}}$. Proposition 1 implies that an adjoint to the forgetful functor F : mod-$\underline{\underline{g}} \to$ mod-$\underline{\underline{h}}$ must take non-zero finite-dimensional $\underline{\underline{h}}$-modules to infinite-dimensional $\underline{\underline{g}}$-modules. Let finmod-$\underline{\underline{g}}$ and finmod-$\underline{\underline{h}}$ denote the categories of finite-dimensional right modules for $\underline{\underline{g}}$ and $\underline{\underline{h}}$. Proposition 1 does not preclude the possibility that the forgetful functor F_0 : finmod-$\underline{\underline{g}} \to$ finmod-$\underline{\underline{h}}$ might have a left or right adjoint. The functor F_0 fails the conditions of the adjoint functor theorem (Mitchell [5], p. 124), so the question of the existence of adjoints to F_0 remains open.

THEOREM 2. *If I_0 is a right adjoint to F_0 :* finmod-$\underline{\underline{g}} \to$ finmod-$\underline{\underline{h}}$*, $W \in$* finmod-$\underline{\underline{h}}$*, $V \in$* finmod-$\underline{\underline{g}}$*, and V is a $U\underline{\underline{g}}$-submodule of* $\hom_{U\underline{\underline{h}}}(U\underline{\underline{g}}, W)$*, then V can be embedded as a $U\underline{\underline{g}}$-submodule in I_0W. Dually, if I_0 is a left adjoint to F_0, $W \in$* finmod-$\underline{\underline{h}}$*, $V \in$* finmod-$\underline{\underline{g}}$*, and V is a quotient module of $W \otimes_{U\underline{\underline{h}}} U\underline{\underline{g}}$, then V is a quotient module of I_0W.*

Thus, if $\hom_{U\underline{\underline{h}}}(U\underline{\underline{g}}, W)$ contains $U\underline{\underline{g}}$-submodules of arbitrarily large finite dimension, so must I_0W, so such an I_0 could not exist. Dually for $W \otimes_{U\underline{\underline{h}}} U\underline{\underline{g}}$ and I_0W.

Proof. Let S : finmod-$\underline{\underline{h}} \to$ mod-$\underline{\underline{h}}$ and T : finmod-$\underline{\underline{g}} \to$ mod-$\underline{\underline{g}}$ denote the inclusion functors. Note that S and T are full and faithful (that is, bijective on morphisms), and that $FT = SF_0$. Suppose that $V \in$ finmod-$\underline{\underline{g}}$ and $W \in$ finmod-$\underline{\underline{h}}$. Then, under the hypotheses of Theorem 2, we have the following composite bijection, natural in V:

$$\begin{array}{ccc}
\text{mod-}\underline{\underline{g}}(TV,\ \hom_{U\underline{\underline{h}}}(U\underline{\underline{g}},\ SW)) & \longrightarrow & \text{mod-}\underline{\underline{h}}(FTV,\ SW) \\
 & & \big\downarrow{\scriptstyle FT\,=\,SF_0} \\
\text{finmod-}\underline{\underline{h}}(F_0V,\ W) & \xleftarrow{\ S^{-1}\ } & \text{mod-}\underline{\underline{h}}(SF_0V,\ SW) \\
\big\downarrow & & \\
\text{finmod-}\underline{\underline{g}}(V,\ I_0W) & &
\end{array}$$

Let us denote this bijection by

$$\eta_V : \text{mod-}\underline{\underline{g}}(TV,\ \hom_{U\underline{\underline{h}}}(U\underline{\underline{g}},\ SW)) \to \text{finmod-}\underline{\underline{g}}(V,\ I_0W) .$$

It is sufficient to show that η_V is monic-preserving. Suppose $u \in \text{mod-}\underline{\underline{g}}(TV,\ \hom_{U\underline{\underline{h}}}(U\underline{\underline{g}},\ SW))$ is monic, that $X \in \text{finmod-}\underline{\underline{g}}$ and $\alpha,\ \beta \in \text{finmod-}\underline{\underline{g}}(X,\ V)$ are such that $\eta_V(u) \circ \alpha = \eta_V(u) \circ \beta$, so that

$$X \underset{\beta}{\overset{\alpha}{\rightrightarrows}} V \xrightarrow{\ \eta_V(u)\ } I_0W$$

commutes. By the naturality of η_V , the following diagram commutes for $\theta \in \{\alpha,\ \beta\}$:

$$\begin{array}{ccc}
\text{mod-}\underline{\underline{g}}(TV,\ \hom_{U\underline{\underline{h}}}(U\underline{\underline{g}},\ SW)) & \xrightarrow{\ \eta_V\ } & \text{finmod-}\underline{\underline{g}}(V,\ I_0W) \\
{\scriptstyle \text{mod-}\underline{\underline{g}}(T\theta,\ 1)}\big\downarrow & & \big\downarrow{\scriptstyle \text{finmod-}\underline{\underline{g}}(\theta,\ 1)} \\
\text{mod-}\underline{\underline{g}}(TX,\ \hom_{U\underline{\underline{h}}}(U\underline{\underline{g}},\ SW)) & \xrightarrow[\ \eta_X\ ]{} & \text{finmod-}\underline{\underline{g}}(X,\ I_0W)
\end{array}$$

Thus $\eta_V(u) \circ \alpha = \eta_X(u \circ T\alpha)$ and $\eta_V(u) \circ \beta = \eta_X(u \circ T\beta)$, so $\eta_X(u \circ T\alpha) = \eta_X(u \circ T\beta)$. It follows that $u \circ T\alpha = u \circ T\beta$ since η_X is bijective. But u is monic, so $T\alpha = T\beta$. T is faithful, so $\alpha = \beta$. This shows $\eta_V(u)$ is a monic in finmod-$\underline{\underline{g}}$. The dual statement is proved similarly. □

Example of the use of Theorem 2. Let k be an arbitrary field. Suppose $\underline{\underline{h}} = \langle h \rangle$, $\underline{\underline{g}} = \langle h,\ g \rangle$, and $[h,\ g] = 0$. Let $W = \langle w \rangle$ be a 1-dimensional $\underline{\underline{h}}$-module such that $w.h = \mu.w$ for some $\mu \in k$. Then $U\underline{\underline{h}}$ is isomorphic to the polynomial algebra in one indeterminate h , and $U\underline{\underline{g}}$ is isomorphic to the polynomial algebra in two commuting indeterminates h and g . An element $f \in \hom_{U\underline{\underline{h}}}(U\underline{\underline{g}},\ W)$ is uniquely determined by its action on the elements g^i , $i \in \{0,\ 1,\ 2,\ 3,\ \ldots\}$, for then $f(g^ih^j) = \mu^jf(g^i)$. Consider the function $f_n \in \hom_{U\underline{\underline{h}}}(U\underline{\underline{g}},\ W)$ defined by

$$f_n(g^i) = \begin{cases} w & \text{for } i = 0, 1, 2, \ldots, n\text{-}1 , \\ 0 & \text{for } i \geq n . \end{cases}$$

We claim that the cyclic $U\underline{g}$-module $f_n.U\underline{g}$ generated by f_n is of dimension n. Indeed, if $k \leq n\text{-}1$,

$$f_n^{g^k}(g^i) = \begin{cases} w & \text{for } i = 0, \ldots, n\text{-}1\text{-}k , \\ 0 & \text{for } i \geq n\text{-}k . \end{cases}$$

Thus a basis for $f_n.U\underline{g}$ is $\left\{f_n, f_n^g, f_n^{g^2}, \ldots, f_n^{g^{n-1}}\right\}$. This means that $\hom_{U\underline{h}}(U\underline{g}, W)$ contains $U\underline{g}$-modules of arbitrarily large finite dimension, and so, by Theorem 2, the forgetful functor F_0 : finmod-$\underline{g}$ → finmod-$\underline{h}$ can not have a right adjoint.

Similar calculations are possible, in principle, in the enveloping algebras of other Lie algebras $\underline{h}$ and $\underline{g}$. However, the multiplication in the enveloping algebra becomes much more involved when $\underline{g}$ is not abelian.

References

[1] J.A. Green, "Axiomatic representation theory for finite groups", *J. Pure Appl. Algebra* **1** (1971), 41-77.

[2] James E. Humphreys, *Introduction to Lie Algebras and Representation Theory* (Graduate Texts in Mathematics, **9**. Springer-Verlag, New York, Heidelberg, Berlin, 1972).

[3] Paul C. Kainen, "Weak adjoint functors", *Math. Z.* **122** (1971), 1-9.

[4] Saunders Mac Lane, *Categories for the Working Mathematician* (Graduate Texts in Mathematics, **5**. Springer-Verlag, New York, Heidelberg, Berlin, 1971).

[5] Barry Mitchell, *Theory of Categories* (Pure and Applied Mathematics, **17**. Academic Press, New York, London, 1965).

[6] David Trushin, "A theorem on induced corepresentations and applications to finite group theory", *J. Algebra* **42** (1976), 173-183.

[7] Nolan R. Wallach, "Induced representations of Lie algebras and a theorem of Borel-Weil", *Trans. Amer. Math. Soc.* **136** (1969), 181-187.

[8] Nolan R. Wallach, "Induced representations of Lie algebras. II", *Proc. Amer. Math. Soc.* **21** (1969), 161-166.

[9] William H. Wilson, "On induced representations of Lie algebras, groups, and coalgebras", submitted.

[10] William H. Wilson, "A functorial version of a construction of Hochschild and Mostow for representations of Lie algebras", *Bull. Austral. Math. Soc.* **18** (1978), 95-98.

Department of Mathematics,
University of Queensland,
St Lucia,
Queensland.

PROC. 18th SRI
CANBERRA 1978, 205-229.

20E10
(20C30)

VARIETIES OF NILPOTENT GROUPS OF SMALL CLASS

L.G. Kovács

1. Introduction

In the dreamtime of the theory of varieties of groups, one might have hoped for the individual knowledge of each variety: for a classification in the strongest sense. The extent to which such hopes have been realized is a remarkable achievement of the subject. R.A. Bryce [7], [8] knows each variety of metabelian groups 'modulo the nilpotent case'. The classification of varieties of nilpotent groups 'of small class' is the subject of this report. Our knowledge in this area comes essentially from Graham Higman's 1965 lecture [12], given to an international conference held here, which dealt with varieties of nilpotent groups of prime exponent p and class less than p. In 1968, M.F. Newman and I presented (in a course of lectures at this University) an extended version of this theory, for varieties of p-power exponent and class less than p, and also for 'torsionfree' nilpotent varieties of arbitrary class. (Our treatment was clumsy, and remained unwritten, but considerable further work is on record in Paul Pentony's thesis [23].) A 1971 paper [14] by A.A. Kljačko (in an extremely inaccessible publication) described yet another version for the case of p-power exponent (and class less than p), apparently independently of Higman's work.

One remarkable aspect of Kljačko's paper was the application of this method to derive information also about certain varieties of p-groups of class *not* less than p. Namely, he established the following

DISTRIBUTIVITY THEOREM. *The lattice of varieties of* p*-power exponent and class at most* c *is distributive if and only if* $c \leq 3$, *or* $c = 4$ *and* $p > 2$, *or* $c = 5$ *and* $p > 5$.

In fact, it was precisely the cases of $c = 4$, $p = 3$ and $c = 5$, $p = 5$ which were still outstanding then. (I must confess that I still can not handle the case $c = 4$, $p = 3$ by this method: Kljačko's paper suppressed the details. I have used *ad hoc* arguments to classify all 3-power exponent varieties of class 4 [unpublished], and found their lattice distributive, in agreement with Kljačko's claim.)

I refer to 'method' with good reason. The situation is so complex that only some qualitative aspect of it can be expressed in any single statement (for example, in the Distributivity Theorem above). On the other hand, while the problem of classifying all nilpotent varieties is theoretically solvable (in an algorithmic sense)*, the approach elucidated by Higman yields a significantly more efficient solution in the small class case, and indeed enables one to prove general statements (instead of having to be content with the knowledge that the proposition at hand is 'decidable'). By general statement I mean not only the Distributivity Theorem, which could be regarded as a case where the decision algorithm fortuitously terminated before we ran out of time: I mean also results like A.G.R. Stewart's theorem [25] that for each c (at least 4) there exist precisely two join-irreducible center-by-metabelian varieties of exponent p and class c (provided $p > c$), or the fact that the variety $\underline{\underline{N}}_c$ of all nilpotent groups of class at most c is generated by $(c-1)$-generator groups but not by $(c-2)$-generator groups (Kovács, Newman, Pentony [16]; see also Levin [18]).

The aim of these lectures is to make 'the method' more accessible. Higman's original [12] is terse to the point of being quite a challenge to read; as a record of a single lecture, it is really just an outline, virtually without proofs, attributions, or references: also, restriction to prime exponent seems worth avoiding today. Nevertheless, it is so rich that I can not cover half his material: I hope the reader will be encouraged, and better prepared, to sample his feast further. Kljačko's [14] is also on the terse side, and as far as I know can not be found in our libraries.

Inevitably, this report will also fail to be self-contained, and there will be many a point where I will wish I had a (better) reference: still, I hope to account for all omissions of non-routine arguments. Instead of attempting to formalize 'the method', I aim to prove two results. One is the Distributivity Theorem (except for

---

* Set one machine to enumerate laws and their consequences: if u is a consequence of v , this will be shown in a finite time. Set another to enumerate finite nilpotent groups and test them for laws: if u is not a consequence of v , a group will turn up to demonstrate this. This does it, for each nilpotent variety is generated by finite groups and definable by a single law.

the case $c = 4$, $p = 3$). The other is also in Kljačko's [14]. For each prime p and positive integer m, let A_p^m denote the dual of the lattice of all subgroups of p-power index in a free abelian group of rank m.

CLASSIFICATION THEOREM. *For $c < p$, the lattice of all varieties of nilpotent groups of p-power exponent and class at most c, is a subdirect product of c lattices, each of which is the direct product of the lattices $A_p^{l(\pi)}$ where π runs through a suitable index set. The index sets and the integers $l(\pi)$ are independent of p* (and will be made explicit in Section 6).

The name 'Classification Theorem' sounds too pretentious for such a result; I use it to suggest that its proof is constructive and would enable us, if we wished, to attach convenient labels to the varieties in question, labels from which one can instantly read off at least some of the most important relationships between these varieties. Unfortunately, I must also acknowledge the incomplete nature of the theorem. For, the phrase 'a subdirect product' hides *the first important open problem* of small class theory: just which subdirect product is it? Thus we do not know precisely which of the available labels would be used. (In the equivalent language of subgroups of free groups: the trouble is that 'the method' only deals conclusively with fully invariant subgroups which lie between successive terms of the lower central series.)

I shall also prove the torsionfree analogues of the two theorems. A variety is called *torsionfree* if it is generated by its torsionfree groups; equivalently, if its free groups are torsionfree. With respect to partial order by set-theoretic inclusion, these varieties form a lattice (which is *not* a sublattice of the lattice of all varieties, for the meet is now the variety generated by the torsionfree groups in the intersection and so can be smaller than the intersection). *The lattice of all torsionfree varieties of nilpotent groups of class at most c is distributive if and only if $c \leq 5$. The Classification Theorem* has the same form as before, with the same index sets and parameters $l(\pi)$, and without any restriction on c; the only change is that A_p^m is replaced by the subspace lattice A_0^m of an m-dimensional rational vector space.

The general case may now be approached as follows. If $\underline{\underline{V}}$ is any variety, it has a well-defined torsionfree core: the variety $\underline{\underline{V}}_0$ generated by the torsionfree groups of $\underline{\underline{V}}$. If $\underline{\underline{V}}$ is nilpotent, it is the join of $\underline{\underline{V}}_0$ and certain varieties $\underline{\underline{V}}_p$, one for each prime p from a finite set, each $\underline{\underline{V}}_p$ of p-power exponent. In a sense, this reduces the study of nilpotent varieties to the torsionfree and prime-power-exponent cases. Of course, when $\underline{\underline{V}}_0$ is trivial, the $\underline{\underline{V}}_p$ are uniquely determined by $\underline{\underline{V}}$ and the reduction is as good as one might wish. However, when $\underline{\underline{V}}_0$ is nonabelian, it does happen that the $\underline{\underline{V}}_p$ are not determined by $\underline{\underline{V}}$, not even if we insist that they be chosen as small as possible and only primes greater than the class of $\underline{\underline{V}}$ occur. The resolution of this difficulty is *the second important open problem* of small class theory.

To conclude this introduction on a more cheerful note, let me draw attention to the unrecorded fact that the torsionfree Classification Theorem leads, via the work of Stewart (*loc. cit*), to such specific results as the following. There are precisely 39 torsionfree varieties of class at most 5 (only one, namely the variety $\underline{\underline{N}}_5 \wedge \underline{\underline{A}}^2$ of all metabelian varieties of class at most 5, failing to lie in, or between successive terms of, the sequence $\underline{\underline{E}}, \underline{\underline{A}}, \underline{\underline{N}}_2, \underline{\underline{N}}_3, \underline{\underline{N}}_4, \underline{\underline{N}}_5$: it is to establish this that Stewart's work is needed here); but there exist infinitely many torsionfree varieties of class 6.

The next six sections contain the technicalities; classification first, distributivity last. Finally, in a postscript I comment on the earlier history of the key ideas. Some of those comments are based on references (included in the list at the end) which only came to my attention after the end of the Institute.

2. Subdirect decompositions

Let F be a noncyclic free group; for convenience, take it to have finite rank (the whole argument would remain valid *mutatis mutandis* without this restriction), and let Y be a free generating set of F. Write the lower central series of F as

$$F = \underline{\underline{N}}_0(F) > F' = \underline{\underline{N}}_1(F) > \ldots > \underline{\underline{N}}_c(F) > \ldots :$$

thus $\underline{\underline{N}}_c(F)$ is the verbal subgroup of F corresponding to the variety $\underline{\underline{N}}_c$ of all nilpotent groups of class (at most) c. As is well known (*cf.* 34.13 in Hanna Neumann's book [22]), if the rank of F is at least c then the lattice of subvarieties of $\underline{\underline{N}}_c$ is dual to the (modular) lattice N_c of fully invariant subgroups of F containing $\underline{\underline{N}}_c(F)$. In this duality, torsionfree subvarieties of $\underline{\underline{N}}_c$ correspond to isolated fully invariant subgroups of F (that is, subgroups U with F/U torsionfree). These subgroups form a lattice N_c^0, in which meet is set intersection and the join $U \vee V$ is obtained by taking $(U \vee V)/UV$ to be the subgroup of F/UV consisting of the elements of finite order (thus N_c^0 is not a sublattice of N_c). It is easy to prove that N_c^0 is modular. For a fixed prime p, the p-power exponent subvarieties of $\underline{\underline{N}}_c$ correspond to fully invariant subgroups of p-power index in F: these form a sublattice N_c^p of N_c. Our subject is therefore the study of the N_c^p and N_c^0. (These lattices do vary with the rank of F when that rank is small, but this dependence will not effect our arguments until the last moment, so for the time being we may ignore it.)

The aim of this section is the following reduction of the problem. Let L_c^0 denote the sublattice $\left\{U \in N_c^0 \,\middle|\, U \leq \underline{\underline{N}}_{c-1}(F)\right\}$ of N_c^0, and L_c^p the sublattice

$$\{U \in N_c \mid U \leq \underline{\underline{N}}_{c-1}(F) \text{ and } |\underline{\underline{N}}_{c-1}(F)/U| \text{ is a power of } p\}$$

of N_c (so L_c^p avoids N_c^p except when $c = 1$). We shall prove that, for $c > 1$, *N_c^0 is always a subdirect product of N_{c-1}^0 and L_c^0, and N_c^p is a subdirect product of N_{c-1}^p and L_c^p provided $c \leq p+1$.* Beyond this section, our time will be devoted to the analysis of the L_c^p and L_c^0, for this reduction (and induction on c) will have established that *N_c^p and N_c^0 are subdirect products of the L_i^p and the L_i (with $1 \leq i \leq c$), respectively.*

Some more comments before we embark on the proof. This reduction is in effect contained in Kljačko's paper [14] for N_c^p and $c < p$ provided F has sufficiently large rank. Our proof needs no restriction on the rank of F. For the Classification Theorem only the case $c < p$ is relevant, and for that the proof we are about to see is really easy. The negative parts of the Distributivity Theorem could be reached via a much simpler version of the reduction, but I can not imagine how the positive part for N_4^3 could be reached by 'the method' without the relevant reduction. The present version of the reduction goes just about as far as possible: when the rank of F is 2, N_4^2 is not a subdirect product of N_3^2 and L_4^2, for the latter are distributive but N_4^2 is not. (In fact, Bryce had shown, in the footnote of page 335 in [7], that N_{p+2}^p is never distributive.)

For the proof, let us write N for $\underline{\underline{N}}_{c-1}(F)$. The key fact is that if $U, V \in N_c^p$ (or N_c^0) then the sublattice of N_c^p (or of N_c^0) generated by U, V, and N, is distributive: assume this for the moment. Then $U \mapsto U \vee N$ and $U \mapsto U \wedge N$ are lattice homomorphisms of N_c^p (or N_c^0) onto N_{c-1}^p (or N_{c-1}^0) and L_c^p (or L_c^0), respectively. The only nontrivial part of this claim is that $N_c^p \to L_c^p$ is onto. To see this, take any W in L_c^p and consider F/W; this is a finitely generated nilpotent group in which the elements of finite order form a finite p-group, namely N/W. Thus F/W is residually a finite p-group (Gruenberg [11], Theorem 2.1 (ii)) and so has a normal subgroup H/W of p-power index which avoids N/W. Take U to be the verbal subgroup of F corresponding to the variety generated by F/H : then $U \in N_c^p$ and $U \wedge N = W$. The subdirect decompositions now follow from the fact that $U \vee N = V \vee N$ and $U \wedge N = V \wedge N$ imply that $U = V$, this implication being valid in every distributive lattice. Instead of memorizing lots of simple results and scattered references, I prefer to keep handy the diagram of the free modular lattice on three generators from which all such claims are easily read off (or disproved):

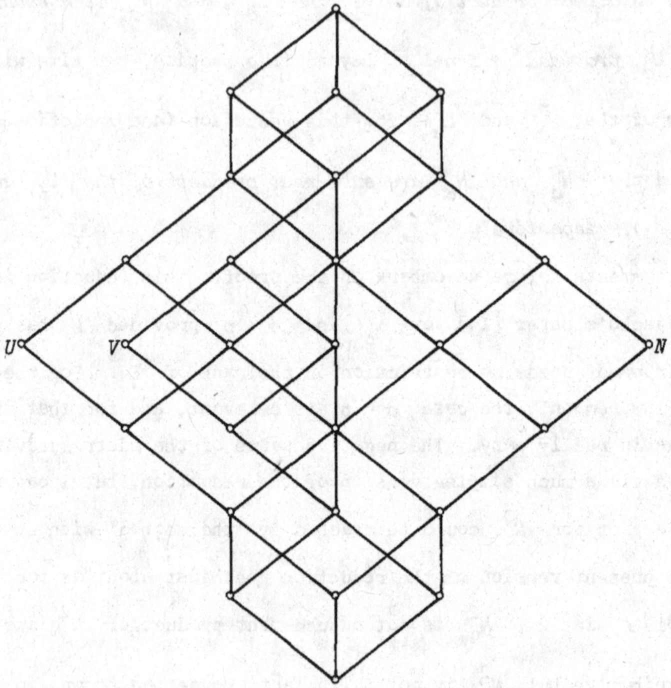

This picture will also start us on our way to deriving a contradiction in case the lattice generated by U, V, and N, is not distributive. Recall that two intervals (pairs of comparable elements), say, $U_1 < U_2$ and $V_1 < V_2$, of a lattice are called perspective if $U_1 = U_2 \wedge V_1$ and $U_2 \vee V_1 = V_2$. Projectivity is then the smallest equivalence relation on the set of all intervals of the lattice such that perspective intervals are projective. If $U_1 < U_2$ and $V_1 < V_2$ are perspective in N_c then by an isomorphism theorem U_2/U_1 and V_2/V_1 are (End F)-isomorphic; if they are perspective in N_c^0 then a finite-index subgroup of U_2/U_1 is (End F)-isomorphic to some finite-index subgroup of V_2/V_1. In these two observations the conclusions are in terms of equivalence relations, hence in the hypotheses perspectivity may be replaced by projectivity. What we need from these comments and the diagram above is that if U, V, N generate a nondistributive sublattice in N_c (or in N_c^0) then $N/\underline{\underline{N}}_c(F)$ and F/N must have (abelian) nontrivial (End F)-isomorphic sections: namely,

$$\frac{(U\vee V)\wedge N}{(U\wedge N)\vee(V\wedge N)} \text{ and } \frac{(U\vee N)\wedge(V\vee N)}{(U\wedge V)\vee N}$$

(which are p-groups) if we are in N_c, or finite-index (torsionfree) subgroups of these if we are in N_c^0.

The next step is to 'refine' this section of F/N according to the lower central series of F/N. This is routine group theory and I omit the details. The conclusion is that, for some i (with $1 \le i < c$), $N/\underline{\underline{N}}_c(F)$ and $\underline{\underline{N}}_{i-1}(F)/\underline{\underline{N}}_i(F)$ have nontrivial (End F)-isomorphic sections which are p-groups if we started in N_c^p and torsionfree if in N_c^0.

At this point, elementary commutator calculus enters the argument. For each positive integer n, consider the endomorphism of F which is defined by $y \mapsto y^n$ for all y in the free generating set Y of F. Induction on j readily yields that this acts as $w \mapsto w^{n^j}$ on every element of (every section of) $\underline{\underline{N}}_{j-1}(F)/\underline{\underline{N}}_j(F)$. The result of the previous paragraph then implies that $n^i \equiv n^c \pmod p$ if we started in N_c^p, or $n^i = n^c$ if in N_c^0, for all n. In the latter case we have the desired contradiction; in the former, only two possibilities remain: either $i = 1$ and $c = p$, or $i = 2$ and $c = p + 1$.

To eliminate the first, consider an endomorphism of F which leaves one element of Y fixed and takes all others to 1. This endomorphism annihilates F' and hence every section of $N/\underline{\underline{N}}_c(F)$; on the other hand we know all fully invariant subgroups of F containing F' and can see that this endomorphism does not annihilate any nontrivial (End F)-admissible section of F/F'.

It is much harder to deal with the second case. Let U_1, U_2, V_1, V_2 be fully invariant subgroups of F such that $\underline{\underline{N}}_2(F) < U_1 < U_2 \le F'$ and $\underline{\underline{N}}_{p+1}(F) < V_1 < V_2 \le \underline{\underline{N}}_p(F)$ while U_2/U_1 and V_2/V_1 are (End F)-isomorphic p-groups: we may as well take these sections to have exponent p. Let G be the subgroup of F generated by any two elements of Y; embed End G in End F by letting every endomorphism of G map all elements of Y outside G to 1, and denote by ε the identity of End G. Then $(U_2/U_1)\varepsilon$ is $(U_2 \cap G)U_1/U_1$ which is (End G)-isomorphic to $(U_2 \cap G)/(U_1 \cap G)$; also $(U_2/U_1)\varepsilon$ is (End G)-isomorphic to $(V_2/V_1)\varepsilon$; so $(U_2 \cap G)/(U_1 \cap G)$ and $(V_2 \cap G)/(V_1 \cap G)$ are (End G)-isomorphic groups of exponent (dividing) p. All fully invariant subgroups of F between F' and $\underline{\underline{N}}_2(F)$ are known: $U_2 = \underline{\underline{B}}_k(F')\underline{\underline{N}}_2(F)$ and $U_1 = \underline{\underline{B}}_{kp}(F')\underline{\underline{N}}_2(F)$ for some k, and so $U_2 \cap G > U_1 \cap G$. (Reminder: $\underline{\underline{B}}_n$ is the variety of all groups of exponent dividing n.) Thus G inherits all the information we need about F and we could work on in G. To simplify notation, we forget G and assume instead that $|Y| = 2$; say, $Y = \{y_1, y_2\}$.

It is time for more commutator calculus. For each positive integer n, we consider the endomorphism of F defined by $y_1 \mapsto y_1^n$, $y_2 \mapsto y_2$. On the cyclic quotient $F'/\underline{\underline{N}}_2(F)$, and hence also on the quotient U_2/U_1 of order p, this endomorphism acts as $u \mapsto u^n$, thus it must act the same way on V_2/V_1. The quotient $\underline{\underline{N}}_p(F)/\underline{\underline{N}}_{p+1}(F)$ has a basis (as free abelian group) consisting of the cosets of the basic commutators of total weight $p+1$ in y_1 and y_2. On the coset b of a basic commutator of weight k in y_1 and $p+1-k$ in y_2, the endomorphism acts as $b \mapsto b^{n^k}$. If V_2/V_1 were (End F)-isomorphic to some section of $[F'' \cap \underline{\underline{N}}_p(F)]\underline{\underline{N}}_{p+1}(F)/\underline{\underline{N}}_{p+1}(F)$, one could exploit the fact that the latter is generated by the cosets of the non-left-normed basic commutators of weight $p+1$, whose weight k in y_1 satisfies $2 \le k \le p-1$, and derive that $n \equiv n^k \pmod{p}$ for some such k and all n, which is clearly impossible. The alternative is that V_2/V_1 is (End F)-isomorphic to a section of $\underline{\underline{N}}_p(F)/[F'' \cap \underline{\underline{N}}_p(F)]\underline{\underline{N}}_{p+1}(F)$ and therefore also to some section, say W_2/W_1, of $F''\underline{\underline{N}}_p(F)/F''\underline{\underline{N}}_{p+1}(F)$. Let p^e denote the exponent of the Sylow p-subgroup of the finitely generated abelian group $F''\underline{\underline{N}}_p(F)/W_1$: then W_2/W_1 is (End F)-isomorphic to a section of

$$\frac{\underline{\underline{B}}_{p^{f-1}}[F''\underline{\underline{N}}_p(F)]F''\underline{\underline{N}}_{p+1}(F)}{\underline{\underline{B}}_{p^f}[F''\underline{\underline{N}}_p(F)]F''\underline{\underline{N}}_{p+1}(F)}$$

for some f with $1 \le f \le e$. The desired contradiction now follows from the fact that these quotients, for all f, are irreducible (End F)-modules of order p^p (while W_2/W_1 has order p). Indeed, a basis $\{b_i \mid 1 \le i \le p\}$ is made up of the cosets of the $[y_2, \underbrace{y_1, \ldots, y_1}_{i}, \underbrace{y_2, \ldots, y_2}_{p-i}]^{p^{f-1}}$, and we shall find it sufficient to contemplate only the action of those endomorphisms φ of F which satisfy

$$y_1\varphi \in y_1^{f_{11}} y_2^{f_{12}} F', \quad y_2\varphi \in y_1^{f_{21}} y_2^{f_{22}} F',$$

$$\det \begin{pmatrix} f_{11} & f_{12} \\ f_{21} & f_{22} \end{pmatrix} \equiv 1 \pmod{p} .$$

These endomorphisms of F act on this basis exactly the way the corresponding elements of $SL(2, p)$ act by homogeneous linear substitutions on the monomials $x^{i-1}y^{p-i}$ in the 2-variable (commutative) polynomial algebra over the field of p elements. The space of homogeneous polynomials of degree $p - 1$ is well known to be an irreducible $SL(2, p)$-module (Brauer and Nesbitt [5]), and our argument is complete.

In the last step, representation theory may be replaced by commutator calculations of Brisley [6]. However, that is a case where 'bare-handed' commutator calculus is pushed near its limits, and we get a clear indication that systematic exploitation of connections with representation theory offers the only hope of further progress. That is precisely what 'the method' does, as we are about to see.

3. Lie representations

The upshot of the previous section is that we are to study certain fully invariant subgroups of a noncyclic free group F which lie between successive terms of the lower central series of F. This is greatly facilitated by a number of shifts in context.

Consider the lower central factors of F as modules for the monoid $\operatorname{End} F$ of all endomorphisms of F : our task is to study their submodules. It is convenient to write these modules additively. As is well known (see G.E. Wall's lectures [28] in this volume as general reference), the restricted direct sum of these modules may be turned into a graded Lie ring $\operatorname{gr} F$ by defining Lie multiplication from group commutator formation, so the action of $\operatorname{End} F$ on $\operatorname{gr} F$ is compatible with this graded Lie ring structure. Thus we have a (monoid) homomorphism from $\operatorname{End} F$ to the monoid $\operatorname{End}(\operatorname{gr} F)$ of all graded Lie ring endomorphisms of $\operatorname{gr} F$, expressing the action of $\operatorname{End} F$ on $\operatorname{gr} F$. Two restriction maps complete a commutative diagram

$$\begin{array}{ccc} \operatorname{End} F & \longrightarrow & \operatorname{End}(\operatorname{gr} F) \\ & \searrow & \downarrow \\ & & \operatorname{End}(F/F') \end{array}$$

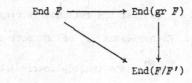

where F/F' is both the commutator factor group of F and the homogeneous component of degree 1 in $\text{gr } F$. Since F is free, each endomorphism of F/F' is the restriction of some endomorphism of F, so this restriction map is onto, and therefore by the diagram so is the other. Because F/F' generates $\text{gr } F$ (see 1.12 in Wall [28]), the vertical restriction map is also one-to-one, hence an isomorphism. It now follows that the horizontal arrow is also onto. The conclusion we want to retain is that the monoid $\text{End}(F/F')$ acts on $\text{gr } F$ as $\text{End}(\text{gr } F)$ and our problem is the study of the $\text{End}(F/F')$-submodules of (the homogeneous components of) $\text{gr } F$. Just before we move on, a warning. As F/F' is abelian, $\text{End}(F/F')$ is usually regarded as a ring; however, its action on $\text{gr } F$ was derived from homomorphisms of mere multiplicative monoids, and it is in fact not compatible with the additive structure of $\text{End}(F/F')$, so $\text{gr } F$ is *not* a module for the *ring* $\text{End}(F/F')$. As a reminder of the need to ignore addition in $\text{End}(F/F')$, one might write $\text{End}^{\times}(F/F')$ instead, but we cannot afford to carry even the notation we have used so far, let alone complicate it: so we shall write simply E.

To prepare for the next shift, let A stand for the ring of all polynomials with integer coefficients in a set of noncommuting indeterminates (the number of indeterminates being the rank of F), and A_c for the additive group of homogeneous polynomials of degree c. Now A_1 is identified with F/F', and then E acts on A_1, and hence on all of A, as the monoid of all linear homogeneous substitutions. Note that each such substitution is a ring endomorphism of A, mapping each A_c into itself. Next, consider the usual Lie multiplication $[a, b] = ab - ba$ which turns A into a graded Lie ring with homogeneous components A_c, and denote by L the Lie subring generated by the indeterminates in A; put $L_c = L \cap A_c$, and note $L_1 = A_1 = F/F'$. Thus L is a graded Lie ring with homogeneous components L_c, and it is clear that L and each L_c admit the given action of E. So far this section has amounted to immediate observations concerning a long string of definitions; by contrast the next point is a deep theorem due to Magnus and Witt (see 3.14 in Wall [28]): the identification of L_1 with F/F' may be extended to an identification of L with $\text{gr } F$. Thus our task becomes the study of E-submodules of the L_c (where by now $E = \text{End}^{\times}(F/F') = \text{End}^{\times}A_1 = \text{End}^{\times}L_1$).

In this task A_c will remain an important aid. Contemplate for a moment: A_c is a free abelian group (with the finite set of all monomials of degree c as basis), so its endomorphism ring $\text{End } A_c$ is just the full matrix ring (of appropriate size) over the ring $\mathbb{Z}$ of integers. Each element e of E acts on A_c as an element, say $e^{\otimes c}$, of $\text{End } A_c$, and $e \mapsto e^{\otimes c}$ is a (multiplicative) homomorphism. The set

$E^{\otimes c}$ of all $e^{\otimes c}$ is a multiplicative submonoid of End A_c, and so the additive subgroup, say E_c, of End A_c generated by $E^{\otimes c}$ is a subring. As additive group, End A_c is free abelian of finite rank, and therefore so is E_c, but beware: $E^{\otimes c}$ is just an additive generating set, not a free basis, for E_c. (I emphasize that the original addition in E remains forgotten, and whatever additive relations there are on $E^{\otimes c}$ are relations in End A_c.) We are after the E_c-submodules of L_c, but the representation theory of the ring E_c is too complicated to approach head on and classification in this generality eludes us.

The final shift will focus on the submodules we really hope to reach: those which are either isolated or of p-power index. Allowing the coefficients of our polynomials to range over the ring $\mathbb{Z}_{(p)}$ of all rational numbers with denominators prime to p (the localization of $\mathbb{Z}$ at p), we obtain a corresponding polynomial algebra $\mathbb{Z}_{(p)}A$ with homogeneous components $\mathbb{Z}_{(p)}A_c$. As the notation suggests, each element of $\mathbb{Z}_{(p)}A_c$ is a product of an element of $\mathbb{Z}_{(p)}$ and an element of A_c. Clearly $\mathbb{Z}_{(p)}A_c$ is a free $\mathbb{Z}_{(p)}$-module with the set of monomials of degree c as basis, and $\mathrm{End}_{\mathbb{Z}_{(p)}} \mathbb{Z}_{(p)}A_c$ is also finitely generated and free as $\mathbb{Z}_{(p)}$-module, embedding End A_c so that $\mathrm{End}_{\mathbb{Z}_{(p)}} \mathbb{Z}_{(p)}A_c = \mathbb{Z}_{(p)}$ End A_c. The $\mathbb{Z}_{(p)}$-submodule $\mathbb{Z}_{(p)}E_c$ is a $\mathbb{Z}_{(p)}$-subalgebra of $\mathrm{End}_{\mathbb{Z}_{(p)}} \mathbb{Z}_{(p)}A_c$. If B is a subgroup of p-power index in L_c, then $\mathbb{Z}_{(p)}B$ is a $\mathbb{Z}_{(p)}$-submodule of finite index in $\mathbb{Z}_{(p)}L_c$ and $L_c \cap \mathbb{Z}_{(p)}B = B$. Conversely, if C is any $\mathbb{Z}_{(p)}$-submodule of finite index in $\mathbb{Z}_{(p)}L_c$ then $L_c \cap C$ is a subgroup of p-power index in L_c and $\mathbb{Z}_{(p)}(L_c \cap C) = C$. If B admits E_c then $\mathbb{Z}_{(p)}B$ admits $\mathbb{Z}_{(p)}E_c$, and if C admits $\mathbb{Z}_{(p)}E_c$ then $L_c \cap C$ admits E_c. Thus $B \mapsto \mathbb{Z}_{(p)}B$ defines a bijection from the set of all E_c-submodules of p-power index in L_c to the set of all $\mathbb{Z}_{(p)}E_c$-submodules of finite index in $\mathbb{Z}_{(p)}L_c$ (it is a bijection because the map defined by $C \mapsto L_c \cap C$ is a twosided inverse for it), and this bijection obviously respects partial order by inclusion. It is an elementary fact that all order-preserving bijections of lattices are lattice isomorphisms. So we may sum up this section: *L_c^p is isomorphic to, and will from now on be thought of as, the lattice of all $\mathbb{Z}_pE_c$-submodules of finite index in* $\mathbb{Z}_{(p)}L_c$. Similarly, *L_c^0 is isomorphic to, and will from now on be thought of as, the lattice of all $\mathbb{Q}E_c$-submodules of* $\mathbb{Q}L_c$, where as usual $\mathbb{Q}$ stands for the field of rational numbers. The point of this last shift is that we shall be able to show that, for $c < p$, $\mathbb{Z}_{(p)}E_c$ is a direct sum of full matrix algebras over $\mathbb{Z}_{(p)}$, and for every case $\mathbb{Q}E_c$ is a direct sum of full matrix algebras over $\mathbb{Q}$, so their representation theory is trivial compared to that of E_c.

4. Symmetric groups

We have reached the point where some detailed calculation can no longer be avoided; please bear with me if I defer motivation for a while. Let X stand for the set of indeterminates in A, and c for the ordered set $\{1, 2, \ldots, c\}$. A monomial $\bar{\mu}$ of degree c is given by choosing c factors from X, in order, with repetitions allowed: that is, by an arbitrary map $\mu : \mathrm{c} \to X$. For the corresponding monomial $\bar{\mu}$ we have $\bar{\mu} = (1\mu)(2\mu) \ldots (c\mu) = \prod_{i \in \mathrm{c}} i\mu$. Let M_c stand for the set of all monomials of degree c. (It is customary to index the elements of X by the first so many natural numbers: this loads the context with an irrelevant order on X and increases typographical complexity so I avoid it, but the reader might prefer to domesticate this section by translation into that traditional form.) Realize the symmetric group S_c as the group of invertible maps σ from c to c, written on the right and composed accordingly. Then S_c also acts on M_c via the composition of maps: $\sigma\bar{\mu} = \overline{\sigma\mu}$ (where $\sigma\mu$ is σ followed by μ from c to c to X). Extend this action linearly to A_c (and also to $\mathbb{Z}_{(p)}\mathrm{A}_c$ and $\mathbb{Q}\mathrm{A}_c$), so A_c becomes a left module for the group ring $\mathbb{Z}S_c$. Let us agree to write endomorphisms of A_c (*qua* abelian group) on the right, and to compose them accordingly: thus A_c is a right $(\mathrm{End}\ \mathrm{A}_c)$-module. However, we are not dealing with a bimodule, because left action of $\mathbb{Z}S_c$ and right action of $\mathrm{End}\ \mathrm{A}_c$ do not commute: in other words, $\mathrm{End}_{\mathbb{Z}S_c}\mathrm{A}_c < \mathrm{End}\ \mathrm{A}_c$. The aim of this section is to show that $\mathrm{End}_{\mathbb{Z}S_c}\mathrm{A}_c \geq E_c$ and, better still, $\mathrm{End}_{\mathbb{Z}_{(p)}S_c}\mathbb{Z}_{(p)}\mathrm{A}_c = \mathbb{Z}_{(p)}E_c$ whenever $p \geq c$, whence of course $\mathrm{End}_{\mathbb{Q}S_c}\mathbb{Q}\mathrm{A}_c = \mathbb{Q}E_c$.

To this end, we shall need to manipulate elements of $\mathrm{End}\ \mathrm{A}_c$. A convenient basis for $\mathrm{End}\ \mathrm{A}_c$ as free abelian group consists of the 'elementary matrices': $e_{\mu\nu}$ takes $\bar{\mu}$ to $\bar{\nu}$, and all monomials other than $\bar{\mu}$ to 0. An element $\sum k_{\mu\nu}e_{\mu\nu}$ of $\mathrm{End}\ \mathrm{A}_c$ commutes on A_c with an element σ of S_c if and only if

$$(\sigma\bar{\kappa}) \sum_{\mu,\nu} k_{\mu\nu}e_{\mu\nu} = \sigma\left(\bar{\kappa} \sum_{\mu,\nu} k_{\mu\nu}e_{\mu\nu}\right),$$

that is,

$$\sum_{\nu} k_{\sigma\kappa,\nu}\bar{\nu} = \sum_{\nu} k_{\kappa\nu}\overline{\sigma\nu} \quad \text{for all } \bar{\kappa} \text{ in } M_c .$$

Comparing coefficients of $\overline{\sigma\nu}$, the last condition amounts to

$$k_{\sigma\kappa,\sigma\nu} = k_{\kappa,\nu} \quad \text{for all } \bar{\kappa}, \bar{\nu} \text{ in } M_c .$$

Consider therefore the action of S_c on the cartesian square $X^C \times X^C$, namely $\sigma : (\mu, \nu) \mapsto (\sigma\mu, \sigma\nu)$: we have that $\sum k_{\mu\nu}e_{\mu\nu} \in \mathrm{End}_{\mathbb{Z}S_c}A_c$ if and only if $k_{\mu\nu}$ is constant on each S_c-orbit in $X^C \times X^C$. Yet another form of this fact is that one basis of $\mathrm{End}_{\mathbb{Z}S_c}A_c$ as free abelian group consists of the "orbit sums" $\sum e_{\mu\nu}$ of elementary matrices where each sum is taken over one complete S_c-orbit of elementary matrices with respect to the action $\sigma : e_{\mu\nu} \mapsto e_{\sigma\mu,\sigma\nu}$.

Consider now an element e of E : it may be described by the matrix of integers $\varepsilon_{x,y}$ $(x, y \in X)$ such that

$$x\varepsilon = \sum_{y\in X} \varepsilon_{x,y}y ,$$

and then $e^{\otimes c}$ acts on M_c as the c-fold Kronecker power:

$$\bar{\mu}e^{\otimes c} = \sum_{\nu} \left(\prod_{i\in C} \varepsilon_{i\mu,i\nu}\right)\bar{\nu} \quad \text{for all } \bar{\mu} \text{ in } M_c ,$$

in other words,

$$\varepsilon^{\otimes c} = \sum_{\mu,\nu} \left(\prod_{i\in C} \varepsilon_{i\mu,i\nu}\right)e_{\mu\nu} .$$

The product $\prod \varepsilon_{i\mu,i\nu}$ does not change when (μ, ν) is replaced by $(\sigma\mu, \sigma\nu)$, for this amounts only to a permutation of its factors which are ordinary integers, so $E^{\otimes c}$ is contained in $\mathrm{End}_{\mathbb{Z}S_c}A_c$ and the easy half of the aim of this section has already been reached.

Towards the hard half, note that the orbit sums $\sum e_{\mu\nu}$ form a basis also for $\mathrm{End}_{\mathbb{Z}_{(p)}S_c}\mathbb{Z}_{(p)}A_c$ as free $\mathbb{Z}_{(p)}$-module. By the easy half, $\mathbb{Z}_{(p)}E_c$ is contained in this module; if it were properly contained, it would lie in some maximal submodule which in turn would be the kernel of a $\mathbb{Z}_{(p)}$-homomorphism onto the unique simple

$\mathbb{Z}_{(p)}$-module $\mathbb{Z}_{(p)}/p\mathbb{Z}_{(p)}$. Thus it suffices to show that if

$$\varphi : \mathrm{End}_{\mathbb{Z}_{(p)}S_c}\mathbb{Z}_{(p)}\mathrm{A}_c \to \mathbb{Z}_{(p)}/p\mathbb{Z}_{(p)}$$

is a $\mathbb{Z}_{(p)}$-homomorphism such that $\varphi(e^{\otimes c}) = 0$ for all $e^{\otimes c}$, then $\varphi = 0$. This is how we shall proceed (adapting the proof of 67.3 in Curtis and Reiner [10]).

The first step is to extend φ to $\mathrm{End}_{\mathbb{Z}_{(p)}}\mathbb{Z}_{(p)}\mathrm{A}_c$ by defining it to map to 0 all but one elementary matrix $e_{\mu\nu}$ from each S_c-orbit: these, together with the orbit sums on which φ is already defined, form another $\mathbb{Z}_{(p)}$-free basis, so this definition is legitimate. Part of the assumption is that

$$\varphi(e^{\otimes c}) = \sum_{\mu,\nu} \varphi(e_{\mu\nu}) \prod_{i\in C} \varepsilon_{i\mu,i\nu} = 0$$

for all $e^{\otimes c}$, that is, for all choices of the $\varepsilon_{x,y}$ in $\mathbb{Z}$. This means that the polynomial

$$\sum_{\mu,\nu} \varphi(e_{\mu\nu}) \prod_{i\in C} z_{i\mu,i\nu}$$

in the set $\{z_{x,y} \mid x, y \in X\}$ of commuting indeterminates vanishes at all substitutions from $\mathbb{Z} + p\mathbb{Z}_{(p)}/p\mathbb{Z}_{(p)}$, that is, from $\mathbb{Z}_{(p)}/p\mathbb{Z}_{(p)}$. Now of course a nonzero polynomial may well vanish at all substitutions from a finite field: but this polynomial is homogeneous of degree c , so as long as $c \leq p$ we can conclude* that i must be the zero polynomial. On the other hand, even the commutative monomials $\prod z_{i\mu,i\nu}$ are different for different S_c-orbits of the pairs (μ, ν) , and for each orbit at most one $\varphi(e_{\mu\nu})$ could be nonzero, so it follows that in fact all $\varphi(e_{\mu\nu})$ are 0 as required.

Thus we have proved that $\mathrm{End}_{\mathbb{Z}S_c}\mathrm{A}_c \geq E_c$ *and* $\mathrm{End}_{\mathbb{Q}S_c}\mathbb{Q}\mathrm{A}_c = \mathbb{Q}E_c$ *for all* c *, while* $\mathrm{End}_{\mathbb{Z}_{(p)}S_c}\mathbb{Z}_{(p)}\mathrm{A}_c = \mathbb{Z}_{(p)}E_c$ *whenever* $c \leq p$. We whall not really need the case $c = p$. (It is easy to see from this argument that the result is sharp, in that the

---

* A nonzero polynomial of degree less than p in each indeterminate cannot vanish at all substitutions from the field of order p : see the proof of Theorem 12.21 in Lausch and Nöbauer [17]. This deals with $c < p$. If $c = p$, use the fact that $z^p - z$ vanishes at all substitutions, so a nonzero homogeneous polynomial of degree p represents the same function as a nonzero polynomial of degree less than p in each indeterminate. (There the argument stops, for $z_1^p z_2 - z_1 z_2^p$ vanishes identically.)

claim fails for $c > p$.)

What makes this result useful is the fact that, *for* $c < p$, *the group algebra* $\mathbb{Z}_{(p)}S_c$ *is a direct sum of full matrix algebras over* $\mathbb{Z}_{(p)}$. The corresponding claim for $\mathbb{Q}S_c$ is better known: but observe that in expressing the relevant elementary matrices as rational linear combinations of elements of S_c , one can get away with denominators which divide $c!$ and are therefore prime to p when $c < p$ (see §5 of Chapter IV in Boerner's text [3]). It will be convenient to exploit this in a separate section.

5. Morita equivalence

This is a pretentious heading, for we need only a very special case of Morita equivalence which is much older: Brauer equivalence might be a more appropriate appellation. For a convenient modern reference, see Chapter 6 in Anderson and Fuller [1]; in particular, Propositions 21.2, 21.7, and Exercises 21.6, 21.5. Let K be a commutative ring with 1 (take $K = \mathbb{Z}_{(p)}$ if that is reassuring, but not if it clouds the issue), m a positive integer, K_m the algebra of $m \times m$ matrices over K. Regard the direct sum $K^{\oplus m}$ of m copies of K first as left K , right K_m module U_m ("row vectors"), then as left K_m , right K module $\overline{U}_m$ ("column vectors"). For the tensor products of these bimodules we have that $U_m \otimes_{K_m} \overline{U}_m \cong K$ and $\overline{U}_m \otimes_K U_m \cong K_m$. It follows that the additive functors $U_m \otimes_{K_m}$ - and $\overline{U}_m \otimes_K$ - provide an equivalence between the categories of left K_m-modules and (left) K-modules (that is, the composites of the two functors are naturally equivalent to the identity functors on these categories). Thus two modules corresponding to each other in this equivalence have isomorphic endomorphism rings and isomorphic submodule lattices.

Suppose now that K is a principal ideal domain and V is any finitely generated left K_m-module such that the product of a nonzero scalar matrix and a nonzero element of V is never zero. Then the corresponding K-module is also finitely generated and "torsionfree", hence a direct sum of, say, n copies of K . It follows that V is the direct sum of n copies of $\overline{U}_m$ as left K_m-module, and the endomorphism ring of V is just K_n (thought of as acting on the right of V). By the same argument with left and right interchanged, a finitely generated "torsionfree" right K_n-module W is a direct sum of, say, l copies of U_n as right K_n-module, and the submodule lattice of W is isomorphic to the submodule lattice of the free K-module of rank l . It is easy to see that in this lattice isomorphism submodules of finite index correspond to submodules of finite index.

It is now a tedious but perfectly elementary exercise to extend these conclusions to the case of direct sums of full matrix rings. Let K remain a principal ideal domain; for π ranging through some finite index set, choose positive integers $m(\pi)$ and let $\oplus K_{m(\pi)}$ be the direct sum of the corresponding matrix rings. Denote by U_π the left K, right $\oplus K_{m(\pi)}$ bimodule obtained from $U_{m(\pi)}$ by letting all the $K_{m(\pi')}$ with $\pi' \neq \pi$ annihilate it, and define $\overline{U}_{m(\pi)}$ similarly. (Note U_π and $\overline{U}_\pi$ depend on π, not just on the integer $m(\pi)$ which may be the same for several distinct elements of the index set.) Let e_π be the identity element of $K_{m(\pi)}$, so $\sum e_\pi$ is the identity element of $\oplus K_{m(\pi)}$. Take any finitely generated left $\oplus K_{m(\pi)}$-module V which is torsionfree in the sense that $\left(k \sum e_\pi\right) v = 0$ with $k \in K$, $v \in V$ implies $k = 0$ or $v = 0$: then V is the direct sum of the $e_\pi V$ and the earlier case can be applied to describe the structure of each $e_\pi V$. The conclusion is that $V = \oplus \overline{U}_\pi^{\oplus n(\pi)}$ for suitable nonnegative integers $n(\pi)$, and the endomorphism ring of V (acting on the right) is $\oplus K_{n(\pi)}$ with π ranging through those indices for which $n(\pi) \neq 0$.

At this stage we are ready to establish the structure of $\mathbb{Z}_{(p)}E_c$ and $\mathbb{Q}E_c$. For we know that $\mathbb{Z}_{(p)}S_c$ has the form $\oplus K_{m(\pi)}$ and $\mathbb{Z}_{(p)}A_c$ is finitely generated and torsionfree: thus $\mathbb{Z}_{(p)}E_c$, which is just the endomorphism ring $\mathrm{End}_{\mathbb{Z}_{(p)}S_c}\mathbb{Z}_{(p)}A_c$, is of the form $\oplus K_{n(\pi)}$; all this with $K = \mathbb{Z}_{(p)}$. The same goes of course for $\mathbb{Q}E_c$, with $K = \mathbb{Q}$. However, there is more to be had from this approach.

To this end, we carry on with extending the comments of the second paragraph of this section. Interchanging right and left, we know that a finitely generated torsionfree right $\oplus K_{n(\pi)}$ module W has the structure $\oplus U_\pi^{\oplus l(\pi)}$ where now the range of π may be smaller than before and the U_π are defined with reference to the $n(\pi)$ rather than the $m(\pi)$. Adapting notation still further, let e_π denote now the identity element of $K_{n(\pi)}$. Every submodule W' of W is of the form $\oplus W'e_\pi$, and W' has finite index in W if and only if each $W'e_\pi$ has finite index in the corresponding We_π. Thus the lattice of finite index submodules of W is the direct product of the lattices of finite index submodules of the We_π. Again, each involves essentially just one matrix ring $K_{n(\pi)}$, so by the previous observation the lattice of finite index submodules of We_π is isomorphic to the lattice of finite index K-submodules of $K^{\oplus l(\pi)}$. In case $K = \mathbb{Z}_{(p)}$, the latter is in turn isomorphic to the lattice $A^p_{l(\pi)}$ of subgroups of p-power index in a free abelian group of rank $l(\pi)$.

We have almost finished the proof of the qualitative part of the Classification Theorem stated in the Introduction. The subdirect decompositions were established in Section 2; we have just proved (read $W = \mathbb{Z}_{(p)}L_c$) that L^p_c is the direct product of the $A^p_{l(\pi)}$ whenever $c < p$; and the same argument with $\mathbb{Q}$ in place of $\mathbb{Z}_{(p)}$, counting all submodules of $\mathbb{Q}L_c$, gives that L^0_c is the direct product of the subspace lattices $A^0_{l(\pi)}$. What remains is the explicit determination of the appropriate range of π and the integers $l(\pi)$. However, the qualitative claims made about these in the Introduction are already at hand. For, from the second paragraph of this section on, all modules and algebras considered have been free as K-modules, so the whole discussion remains invariant as we change K from $\mathbb{Z}_{(p)}$ to $\mathbb{Q}$ or even all the way to the complex field $\mathbb{C}$ - neither the range of π nor the multiplicities $l(\pi)$ change in the process. Thus indeed they are independent of the prime p: all they depend on is the class c and that long forgotten parameter, the rank of F. The quantitative details will be given in the next section.

6. The multiplicity formula

Let r denote the common rank of F and A. The quantitative details needed to complete the Classification Theorem have been worked out long ago, in the context of the representation theory of the general linear groups $GL(r, \mathbb{C})$. Put $G = GL(r, \mathbb{C})$: this also acts on $\mathbb{C}A_c$, by invertible linear homogeneous substitutions with complex coefficients, and the Kronecker powers $g^{\otimes c}$ (for $g \in G$) also span $\mathrm{End}_{\mathbb{C}S_c} \mathbb{C}A_c$, that is, $\mathbb{C}E_c$: for a qualitative description, see §67 in Curtis and Reiner [10]; for a wealth of detail, I find Boerner [3] the most readable source. What we need here is that the complex irreducible representations of S_c are traditionally indexed by the partitions π of c, so initially the set of all these is our range for π. There are formulas for the $m(\pi)$, but we don't need those here. The irreducibles which occur with positive multiplicity $n(\pi)$ in $\mathbb{C}A_c$ are precisely those labelled with partitions into at most r parts, so the set of these becomes our final range for π, relevant in describing $\mathbb{Z}_{(p)}E_c$ and $\mathbb{Q}E_c$. Again, there are formulas for the dimensions $n(\pi)$, but we don't need them either. The character of G afforded by $\mathbb{C}L_c$ was, I believe, first given by Angeline Brandt [4], and the corresponding formula for the multiplicities $l(\pi)$ follows so directly that it must have been known to her, though the earliest that I can find it in print is in Wever's paper [29]. It is

$$l(\pi) = \frac{1}{c} \sum_{d|c} \mu(d) \chi_\pi (\sigma^{c/d})$$

where μ is the Möbius function, χ_π is the irreducible character of S_c indexed by π, and σ is the cyclic permutation $(12 \dots c)$. There are various ways of evaluating $\chi_\pi(\sigma^{c/d})$; let me give one, just for flavour. If the parts of π are $k_1, \dots, k_s$ so that $k_1 \geq k_2 \geq \dots \geq k_s$ (and $k_1 + k_2 + \dots + k_s = c$), then $\chi_\pi(\sigma^{c/d})$ is the alternating sum of the multinomial coefficients

$$\left\{\begin{matrix} c/d \\ \dfrac{k_1+1\tau-1}{d}, \dfrac{k_2+2\tau-2}{d}, \dots, \dfrac{k_s+s\tau-s}{d} \end{matrix}\right\}$$

over all permutations τ of $\{1, 2, \dots, s\}$, with sign according to the parity of τ, and subject to the convention that the multinomial coefficient is 0 unless all its entries are nonnegative integers. At least, this is how I read pp. 134-135 in Murnaghan's book [21].

As I said, the multiplicity formula is obtained from the character formula. That, in turn, may be easily derived (as in Wever [30]) from the second of 'Witt's formulas' (reporduced as Theorem 5.11 in Magnus, Karrass, Solitar [20], for instance), although Brandt preferred a different argument. The proof in Kljačko's [14] left as exercise for the reader a step which seems every bit as hard as the multiplicity formula itself, but in a more recent paper [15] he indicates a nice proof and gives a fascinating interpretation - paraphrased via Frobenius reciprocity, this reads as follows. Take any faithful linear (complex) character of the cyclic group generated by σ, and induce it to S_c: then $l(\pi)$ is also the multiplicity of χ_π in this induced character. He deduces that, when $c > 6$, the only partitions π with $l(\pi) = 0$ are those corresponding to the two linear characters of S_c. (This was conjectured in Pentony's unpublished thesis [23].) The multiplicities for $c \leq 6$ are as follows (from Thrall [27] who used a recursive method and tabulated l for $c \leq 10$; but see Brandt's [4] for a correction in the case $c = 10$ itself):

the value of l is	0	1	2	3
at the partitions		1		
	2	1^2		
	$3, 1^3$	21		
	$4, 2^2, 1^4$	$31, 21^2$		
	$5, 1^5$	$41, 32, 31^2, 2^21, 21^3$		
	$6, 2^3, 1^6$	$51, 42, 3^2, 31^3, 21^4$	$41^2, 2^21^2$	321

where the usual notation for partitions has been used: for example, 41^2 stands for the partition $4 + 1 + 1$ of 6.

7. Large class

Having completed the Classification Theorems, there is relatively little left to establish the Distributivity Theorems. The fact that $l(\pi) \leq 1$ whenever $c \leq 5$ but $l(\pi) > 1$ for some π when $c = 6$ means that the torsionfree Classification Theorem implies the torsionfree Distributivity Theorem, and the same happens for the p-power exponent case whenever $p \geq 7$. The case of $c \leq 3$ has been covered by Jónsson [13] and Remeslennikov [24]. As I have already mentioned, Bryce [7] has demonstrated the nondistributivity of N^p_{p+2} for all primes p, exploiting the breakdown of the subdirect reduction discussed in our Section 2. This leaves the cases N^3_4, N^5_4, N^5_5. For N^3_4 I need *ad hoc* methods which are not suitable for presentation here. The distributivity of N^5_4 follows from the Classification Theorem. What is left then is N^5_5; by Section 2, it will suffice to prove that L^5_5 is not distributive. (Kljačko [14] proves the nondistributivity of L^2_4, a different route to N^2_4 from Bryce's. The argument is exactly analogous to that which I will present for L^5_5. The case of N^2_4 was also dealt with by Belov [2].)

Before embarking on the discussion of this "large class" case, some general comments. As observed before the final shift in Section 3, the elusive general problem is the analysis of the E_c-submodules of L_c. We have proved that $\mathbb{Q}E_c$ is a direct sum of full matrix rings over $\mathbb{Q}$, and observed that additively E_c is free abelian of finite rank: so E_c is a $\mathbb{Z}$-order in the excellent algebra $\mathbb{Q}E_c$. What we have done for primes with $p > c$ may be viewed as describing the localizations of

E_c (and L_c) at these primes, standard steps in the investigation of any $\mathbb{Z}$-order. The main step of Section 4 can be reinterpreted to say that E_c is contained in another, more tractable, $\mathbb{Z}$-order in the same algebra, namely in $\mathrm{End}_{\mathbb{Z}S_c} A_c$, and that the finite index of E_c in $\mathrm{End}_{\mathbb{Z}S_c} A_c$ is divisible only by primes strictly less than c. It is useful to know that L_c admits the action of this larger $\mathbb{Z}$-order (for $L_c = \Omega_c A_c$ where Ω_c is a suitable element of $\mathbb{Z}S_c$, whose introduction was attributed to Otto Grün by Magnus in [19]: see Theorems 5.16, 5.17 in Magnus, Karrass Solitar [20]. For example, the lattice of $(\mathrm{End}_{\mathbb{Z}S_c} A_c)$-submodules of L_c is a sublattice of the lattice of E_c-submodules, and in aiming for a nondistributivity result it is sufficient if one can succeed in that sublattice. (This help is not needed in dealing with L_5^5, for then $p \geq c$, but it does matter in the case of L_4^2.) The point is that we have more information on $\mathbb{Z}S_c$ (and hence also on $\mathrm{End}_{\mathbb{Z}S_c} A_c$ even in the context of small primes, and this can be exploited to good effect.

As the situation is now tighter, for comfort let us assume that the rank r is large. Then there exist one-to-one maps $\mu : C \to X$, and the corresponding monomials $\overline{\mu}$ have trivial stabilizers in S_c, so S_c acts regularly on the orbit of such a $\overline{\mu}$. What we need is that A_c has a direct summand, namely $\mathbb{Z}S_c\overline{\mu}$, which is a regular $\mathbb{Z}S_c$-module. (This makes it particularly easy to see that $\mathrm{End}_{E_c} A_c$ is just the image of $\mathbb{Z}S_c$ in $\mathrm{End}_{\mathbb{Z}} A_c$: a potentially useful fact, but irrelevant to our immediate purpose.)

To come to the point, let us take $p = c = 5 \leq r$ and $\mu : C \to X$ one-to-one so $\mathbb{Z}_{(5)}S_5\overline{\mu}$ is a regular direct summand of $\mathbb{Z}_{(5)}A_5$, and recall from Section 4 that in th case $\mathrm{End}_{\mathbb{Z}_{(5)}S_5} A_5 = \mathbb{Z}_{(5)}E_5$. If W is an indecomposable direct summand of $\mathbb{Z}_{(5)}S_5\overline{\mu}$ (*qua* $\mathbb{Z}_{(5)}S_5$-module), then it is also a direct summand of $\mathbb{Z}_{(5)}A_5$; equivalently, $\mathbb{Z}_{(5)}A_5$ has an idempotent $\mathbb{Z}_{(5)}S_5$-endomorphism, say f, with $\mathbb{Z}_{(5)}A_5 f = W$. The indecomposability of W means that f is primitive in $\mathbb{Z}_{(5)}E_5$. When $\mathbb{Q}W$ decompose as a direct sum, $\oplus \overline{U}_\pi^{\oplus t(\pi)}$ say, of irreducible $\mathbb{Q}S_c$-modules, this corresponds to f being no longer primitive in $\mathbb{Q}E_5$: instead, $f = \sum_\pi \sum_{i=1}^{t(\pi)} f_{\pi,i}$ with the $f_{\pi,i}$ pair-wise orthogonal idempotents primitive in $\mathbb{Q}E_5$ and $\mathbb{Q}Wf_{\pi,i} \cong \overline{U}_\pi$. The way we labelled the simple components of $\mathbb{Q}E_5$ by partitions means, in this context, that $f_{\pi,1}, \ldots, f_{\pi,t(\pi)}$ are in the simple component labelled by π. In exact parallel, of course, $f\mathbb{Z}_{(5)}E_5$ is an indecomposable right ideal in $\mathbb{Z}_{(5)}E_5$ but $f\mathbb{Q}E_5 = \oplus\oplus f_{\pi,i}\mathbb{Q}E_5$, with $f_{\pi,i}\mathbb{Q}E_5$ belonging to the isomorphism type of irreducible $\mathbb{Q}E_5$-modules labelled by π. As W was an indecomposable direct summand of the regular $\mathbb{Z}_{(5)}S_5$-module $\mathbb{Z}_{(5)}S_5\bar{\mu}$, we know that the $t(\pi)$ form a column in the decomposition matrix (see the beginning of the proof of 83.9 in Curtis and Reiner [10], including the comment in the footnote which enables one to avoid completion given that $\mathbb{Q}$ is a splitting field for S_5); and we are still free to choose W to obtain whichever column we like. (I shall not reproduce the decomposition matrix here; it is not hard to calculate.) Let us choose W so we get the column corresponding to the 3-dimensional composition factor of the permutation representation of degree 5 over the field of 5 elements: then $t(41) = t(31^2) = 1$ (and t vanishes at all other partitions). Since we also have $l(41) = l(31^2) = 1$, in this case $f\mathbb{Q}E_5$ is isomorphic to a submodule of $\mathbb{Q}L_5$, namely to a submodule which is a direct sum of two irreducibles, say, of U and V. What we need is that $f\mathbb{Q}E_5$ has homomorphisms onto each of U and V, say, $\alpha : f\mathbb{Q}E_5 \twoheadrightarrow U$ and $\beta : f\mathbb{Q}E_5 \twoheadrightarrow V$. Now $U = \mathbb{Q}\alpha(f\mathbb{Z}_{(5)}E_5) = \mathbb{Q}(U \cap \mathbb{Z}_{(5)}L_5)$ and so for some power of 5, say 5^a, we have $0 \neq 5^a\alpha(f\mathbb{Z}_{(5)}E_5) \leq U \cap \mathbb{Z}_{(5)}L_5$; similarly, $0 \neq 5^b\beta(f\mathbb{Z}_{(5)}E_5) \leq V \cap \mathbb{Z}_{(5)}L_5$ for some b. All we need of this is that $f\mathbb{Z}_{(5)}E_5$ has two disjoint nonzero homomorphic images in $\mathbb{Z}_{(5)}L_5$: for brevity, let us call these U' and V'. The last point is that, because f is primitive in $\mathbb{Z}_{(5)}E_5$, all proper submodules of $f\mathbb{Z}_{(5)}E_5$ are contained in a unique maximal submodule. (I find this rather awkward to dig out of Curtis and Reiner [10]. For a start, as $\mathbb{Q}E_5$ is a direct sum of full matrix rings over $\mathbb{Q}$, the proof of 76.29 gives that $f\mathbb{Z}_{(5)}E_5$ remains indecomposable after 5-adic completion, so f remains primitive; then a standard result on lifting idempotents, 77.10, yields that f is primitive modulo 5. Thus modulo 5 we get that $f\mathbb{Z}_{(5)}E_5$ becomes a principal indecomposable module for the finite image of $\mathbb{Z}_{(5)}E_5$, and hence by 54.11 has a unique maximal submodule. The preimage of this modulo 5 will do, by Nakayama's Lemma.) This maximal submodule then contains the kernels of the homomorphisms onto U' and V'; let U'' and V'' be the images of the maximal submodule in U' and V', respectively. Now U'/U'' and V'/V'' are both isomorphic to the unique simple homomorphic image of

$f\mathbb{Z}_{(5)}E_5$. Recall that $U \oplus V$ was a direct summand of $\mathbb{Q}L_5$; let C be any direct complement, so $U \oplus V \oplus C = \mathbb{Q}L_5$, and put $C' = C \cap \mathbb{Z}_{(5)}L_5$. Then $U'' \oplus V'' \oplus C'$ has finite index in $\mathbb{Z}_{(5)}L_c$, and the quotient $(U' \oplus V' \oplus C')/(U'' \oplus V'' \oplus C')$ is a direct sum of two isomorphic summands, namely, of $(U' \oplus V'' \oplus C')/(U'' \oplus V'' \oplus C')$ and $(U'' \oplus V' \oplus C')/(U'' \oplus V'' \oplus C')$. Thus these two summands and their 'diagonal' violate the distributive law, proving that L_5^5 is not distributive. This completes the proof.

8. Postscript

All the background for this was available by the late 1930's: the Magnus-Witt Theorem, and enough of Brauer's theory of modular representations (including his observation that results from Schur's dissertation concerning representations of general linear groups on tensor spaces remain valid in finite characteristic for the small degree case). The first mention of Grün's Ω_c was in a lecture [19] given by Magnus to a week-long group theory meeting at Göttingen in June 1939 (*Crelle* devoted a whole issue to the proceedings): Magnus drew attention to the problem of investigating the action of homogeneous linear substitutions on homogeneous components of free Lie rings, and to the relevance of this in the study of fully invariant subgroups. There are indications that not only Grün and Magnus but also Witt and Zassenhaus were using such ideas at the time, though I have found no evidence for Higman's guess [12] that Witt might have been in possession of the character formula. On the other side of the Atlantic, Thrall got very much closer to the developments reported on here. In his paper [26] (which was submitted before *Crelle*'s Göttingen issue appeared), he used Lie representations systematically for determining all characteristic subgroups in the last term of the lower central series of free groups of $\underline{\underline{B}}_p \wedge \underline{\underline{N}}_c$ for $c < p$, and referred also to the p-power exponent case. Presumably with this motivation, he proceeded with a systematic study of Lie representations in [27], and this was carried on by Brandt in [4]. In the late 1940's Wever took the matter further in several papers, but his applications concerned specific fully invariant subgroups rather than general classification, and interest in Lie representations favoured one-dimensional submodules ("invariants"), perhaps on account of a similar emphasis in Magnus [19]. When variety theory came to life again in the 1950's, it seemed to have no contact with these efforts. Even after Magnus had drawn Hanna Neumann's attention to the relevance of Burrow's then still recent work [9] on 'Lie invariants' (see page 104 in [22]), we did not catch on. From our point of view it did not help to focus on invariants - this seems to have led to the incorrect conjecture expressed in Problem 14 of [22] and, by making the result plausible, encouraged the oversight in 35.35 of [22]. Still, we had little excuse for being as stunned as we were by Graham Higman's lecture [12] which finally opened our eyes.

Higman's account [12] is in terms of prime characteristic. Kljačko [14] worked with p-adic completions (even to the point of starting with a free pro-p-group). Newman and I used localization at p (Mal'cev completions of free nilpotent groups). The present approach is closest to that envisaged in the closing paragraph of Pentony's thesis [23]; it developed in the course of writing up this paper, and (as well as including more detail) deviates substantially from what I actually said in the lectures. In allowing one to view much of the work as a study of the $\mathbb{Z}$-order E_c, it may point in the direction one could proceed beyond the present boundaries.

I am indebted to several participants of the Institute for long and helpful discussions; particularly to P. Fitzpatrick, M.F. Newman, M.G. Schooneveldt, and G. E. Wall.

References

[1] Frank W. Anderson, Kent R. Fuller, *Rings and Categories of Modules* (Graduate Texts in Mathematics, 13. Springer-Verlag, New York, Heidelberg, Berlin, 1974).

[2] Ю.А. Белов [Ju.A. Belov], "К вопросу а решетке нильпотентных многообразий групп класса 4" [On the question of the lattice of nilpotent varieties of groups of class 4], *Algebra i Logika* 9 (1970), 623-628; *Algebra and Logic* 9 (1970), 371-374.

[3] Hermann Boerner, *Representations of Groups* (North-Holland, Amsterdam, 1963).

[4] Angeline J. Brandt, "The free Lie ring and Lie representations of the full linear group", *Trans. Amer. Math. Soc.* 56 (1944), 528-536.

[5] R. Brauer and C. Nesbitt, "On the modular characters of groups", *Ann. of Math.* (2) 42 (1941), 556-590.

[6] Warren Brisley, "Varieties of metabelian p-groups of class p, $p+1$ ", *J. Austral. Math. Soc.* 12 (1971), 53-62.

[7] R.A. Bryce, "Metabelian groups and varieties", *Philos. Trans. Roy. Soc. London Ser. A* 266 (1970), 281-355.

[8] R.A. Bryce, "Varieties of metabelian p-groups", *J. London Math. Soc.* (2) 13 (1976), 363-380.

[9] Martin D. Burrow, "Invariants of free Lie rings", *Comm. Pure Appl. Math.* 11 (1958), 419-431.

[10] Charles W. Curtis, Irving Reiner, *Representation Theory of Finite Groups and Associative Algebras* (Pure and Applied Mathematics, 11. Interscience [John Wiley & Sons], New York, London, 1962).

[11] K.W. Gruenberg, "Residual properties of infinite soluble groups", *Proc. London Math. Soc.* (3) 7 (1957), 29-62.

[12] Graham Higman, "Representations of general linear groups and varieties of p-groups", *Proc. Internat. Conf. Theory of Groups,* Canberra, 1965, 167-173 (Gordon and Breach, New York, London, Paris, 1967).

[13] Bjarni Jónsson, "Varieties of groups of nilpotency three", *Notices Amer. Math. Soc.* 13 (1966), 488.

[14] A.A. Клячко [A.A. Kljačko], "Многообразия p-групп малого класса" [Varieties of p-groups of a small class], *Ordered Sets and Lattices* No.1, 31-42 (Izdat. Saratov Univ., Saratov, 1971).

[15] A.A. Клячко [A.A. Kljačko], "Элементы Ли в тензорной алгебре" [Lie elements in a tensor algebra], *Sibirsk. Mat. Ž.* 15 (1974), 1296-1304, 1430; *Siberian Math. J.* 15 (1974), 914-921.

[16] L.G. Kovács, M.F. Newman and P.F. Pentony, "Generating groups of nilpotent varieties", *Bull. Amer. Math. Soc.* 74 (1968), 968-971.

[17] Hans Lausch and Wilfred Nöbauer, *Algebra of Polynomials* (North-Holland Mathematical Library, 5. North-Holland, Amsterdam, London; American Elsevier, New York, 1973).

[18] Frank Levin, "Generating groups for nilpotent varieties", *J. Austral. Math. Soc.* 11 (1970), 28-32; Corrigendum, *ibid.* 12 (1971), 256.

[19] Wilhelm Magnus, "Über Gruppen und zugeordnete Liesche Ringe", *J. reine angew. Math.* 182 (1940), 142-149.

[20] Wilhelm Magnus, Abraham Karrass, Donald Solitar, *Combinatorial Group Theory: Presentations of groups in terms of generators and relations* (Pure and Appl. Math. 13. Interscience [John Wiley & Sons], New York, London, Sydney, 1966).

[21] Francis D. Murnaghan, *The Theory of Group Representations* (The Johns Hopkins Press, Baltimore, 1938).

[22] Hanna Neumann, *Varieties of Groups* (Ergebnisse der Mathematik und ihrer Grenzgebiete, 37. Springer-Verlag, Berlin, Heidelberg, New York, 1967).

[23] Paul Pentony, "Laws in torsion-free nilpotent varieties with particular reference to the laws of free nilpotent groups" (PhD thesis, Australian National University, Canberra, 1970. See also: Abstract: *Bull. Austral. Math. Soc.* 5 (1971), 283-284).

[24] В.Н. Ремесленников [V. N. Remeslennikov], "Два замечания о 3-ступенно нильпотентных группах" [Two remarks on 3-step nilpotent groups], *Algebra i Logika Sem.* 4 (1965), no. 2, 59-65.

[25] A.G.R. Stewart, "On centre-extended-by-metabelian groups", *Math. Ann.* 185 (1970), 285-302.

[26] Robert M. Thrall, "A note on a theorem by Witt", *Bull. Amer. Math. Soc.* 47 (1941), 303-308.

[27] R.M. Thrall, "On symmetrized Kronecker powers and the structure of the free Lie ring", *Amer. J. Math.* 64 (1942), 371-388.

[28] G.E. Wall, "Lie methods in group theory", these proceedings, 137-173.

[29] Franz Wever, "Operatoren in Lieschen Ringen", *J. reine angew. Math.* 187 (1949), 44-55.

[30] Franz Wever, "Über Invarianten in Lie'schen Ringen", *Math. Ann.* 120 (1949), 563-580.

Department of Mathematics,
Institute of Advanced Studies,
Australian National University,
Canberra, ACT.

Vol. 521: G. Cherlin, Model Theoretic Algebra – Selected Topics. IV, 234 pages. 1976.

Vol. 522: C. O. Bloom and N. D. Kazarinoff, Short Wave Radiation Problems in Inhomogeneous Media: Asymptotic Solutions. V. 104 pages. 1976.

Vol. 523: S. A. Albeverio and R. J. Høegh-Krohn, Mathematical Theory of Feynman Path Integrals. IV, 139 pages. 1976.

Vol. 524: Séminaire Pierre Lelong (Analyse) Année 1974/75. Edité par P. Lelong. V, 222 pages. 1976.

Vol. 525: Structural Stability, the Theory of Catastrophes, and Applications in the Sciences. Proceedings 1975. Edited by P. Hilton. VI, 408 pages. 1976.

Vol. 526: Probability in Banach Spaces. Proceedings 1975. Edited by A. Beck. VI, 290 pages. 1976.

Vol. 527: M. Denker, Ch. Grillenberger, and K. Sigmund, Ergodic Theory on Compact Spaces. IV, 360 pages. 1976.

Vol. 528: J. E. Humphreys, Ordinary and Modular Representations of Chevalley Groups. III, 127 pages. 1976.

Vol. 529: J. Grandell, Doubly Stochastic Poisson Processes. X, 234 pages. 1976.

Vol. 530: S. S. Gelbart, Weil's Representation and the Spectrum of the Metaplectic Group. VII, 140 pages. 1976.

Vol. 531: Y.-C. Wong, The Topology of Uniform Convergence on Order-Bounded Sets. VI, 163 pages. 1976.

Vol. 532: Théorie Ergodique. Proceedings 1973/1974. Edité par J.-P. Conze and M. S. Keane. VIII, 227 pages. 1976.

Vol. 533: F. R. Cohen, T. J. Lada, and J. P. May, The Homology of Iterated Loop Spaces. IX, 490 pages. 1976.

Vol. 534: C. Preston, Random Fields. V, 200 pages. 1976.

Vol. 535: Singularités d'Applications Differentiables. Plans-sur-Bex. 1975. Edité par O. Burlet et F. Ronga. V, 253 pages. 1976.

Vol. 536: W. M. Schmidt, Equations over Finite Fields. An Elementary Approach. IX, 267 pages. 1976.

Vol. 537: Set Theory and Hierarchy Theory. Bierutowice, Poland 1975. A Memorial Tribute to Andrzej Mostowski. Edited by W. Marek, M. Srebrny and A. Zarach. XIII, 345 pages. 1976.

Vol. 538: G. Fischer, Complex Analytic Geometry. VII, 201 pages. 1976.

Vol. 539: A. Badrikian, J. F. C. Kingman et J. Kuelbs, Ecole d'Eté de Probabilités de Saint Flour V-1975. Edité par P.-L. Hennequin. IX, 314 pages. 1976.

Vol. 540: Categorical Topology, Proceedings 1975. Edited by E. Binz and H. Herrlich. XV, 719 pages. 1976.

Vol. 541: Measure Theory, Oberwolfach 1975. Proceedings. Edited by A. Bellow and D. Kölzow. XIV, 430 pages. 1976.

Vol. 542: D. A. Edwards and H. M. Hastings, Čech and Steenrod Homotopy Theories with Applications to Geometric Topology. VII, 296 pages. 1976.

Vol. 543: Nonlinear Operators and the Calculus of Variations, Bruxelles 1975. Edited by J. P. Gossez, E. J. Lami Dozo, J. Mawhin, and L. Waelbroeck, VII, 237 pages. 1976.

Vol. 544: Robert P. Langlands, On the Functional Equations Satisfied by Eisenstein Series. VII, 337 pages. 1976.

Vol. 545: Noncommutative Ring Theory. Kent State 1975. Edited by J. H. Cozzens and F. L. Sandomierski. V, 212 pages. 1976.

Vol. 546: K. Mahler, Lectures on Transcendental Numbers. Edited and Completed by B. Diviš and W. J. Le Veque. XXI, 254 pages. 1976.

Vol. 547: A. Mukherjea and N. A. Tserpes, Measures on Topological Semigroups: Convolution Products and Random Walks. V, 197 pages. 1976.

Vol. 548: D. A. Hejhal, The Selberg Trace Formula for PSL (2,ℝ). Volume I. VI, 516 pages. 1976.

Vol. 549: Brauer Groups, Evanston 1975. Proceedings. Edited by D. Zelinsky. V, 187 pages. 1976.

Vol. 550: Proceedings of the Third Japan – USSR Symposium on Probability Theory. Edited by G. Maruyama and J. V. Prokhorov. VI, 722 pages. 1976.

Vol. 551: Algebraic K-Theory, Evanston 1976. Proceedings. Edited by M. R. Stein. XI, 409 pages. 1976.

Vol. 552: C. G. Gibson, K. Wirthmüller, A. A. du Plessis and E. J. N. Looijenga. Topological Stability of Smooth Mappings. V, 155 pages. 1976.

Vol. 553: M. Petrich, Categories of Algebraic Systems. Vector and Projective Spaces, Semigroups, Rings and Lattices. VIII, 217 pages. 1976.

Vol. 554: J. D. H. Smith, Mal'cev Varieties. VIII, 158 pages. 1976.

Vol. 555: M. Ishida, The Genus Fields of Algebraic Number Fields. VII, 116 pages. 1976.

Vol. 556: Approximation Theory. Bonn 1976. Proceedings. Edited by R. Schaback and K. Scherer. VII, 466 pages. 1976.

Vol. 557: W. Iberkleid and T. Petrie, Smooth S^1 Manifolds. III, 163 pages. 1976.

Vol. 558: B. Weisfeiler, On Construction and Identification of Graphs. XIV, 237 pages. 1976.

Vol. 559: J.-P. Caubet, Le Mouvement Brownien Relativiste. IX, 212 pages. 1976.

Vol. 560: Combinatorial Mathematics, IV, Proceedings 1975. Edited by L. R. A. Casse and W. D. Wallis. VII, 249 pages. 1976.

Vol. 561: Function Theoretic Methods for Partial Differential Equations. Darmstadt 1976. Proceedings. Edited by V. E. Meister, N. Weck and W. L. Wendland. XVIII, 520 pages. 1976.

Vol. 562: R. W. Goodman, Nilpotent Lie Groups: Structure and Applications to Analysis. X, 210 pages. 1976.

Vol. 563: Séminaire de Théorie du Potentiel. Paris, No. 2. Proceedings 1975–1976. Edited by F. Hirsch and G. Mokobodzki. VI, 292 pages. 1976.

Vol. 564: Ordinary and Partial Differential Equations, Dundee 1976. Proceedings. Edited by W. N. Everitt and B. D. Sleeman. XVIII, 551 pages. 1976.

Vol. 565: Turbulence and Navier Stokes Equations. Proceedings 1975. Edited by R. Temam. IX, 194 pages. 1976.

Vol. 566: Empirical Distributions and Processes. Oberwolfach 1976. Proceedings. Edited by P. Gaenssler and P. Révész. VII, 146 pages. 1976.

Vol. 567: Séminaire Bourbaki vol. 1975/76. Exposés 471–488. IV, 303 pages. 1977.

Vol. 568: R. E. Gaines and J. L. Mawhin,Coincidence Degree, and Nonlinear Differential Equations. V, 262 pages. 1977.

Vol. 569: Cohomologie Etale SGA 4½. Séminaire de Géométrie Algébrique du Bois-Marie. Edité par P. Deligne. V, 312 pages. 1977.

Vol. 570: Differential Geometrical Methods in Mathematical Physics, Bonn 1975. Proceedings. Edited by K. Bleuler and A. Reetz. VIII, 576 pages. 1977.

Vol. 571: Constructive Theory of Functions of Several Variables, Oberwolfach 1976. Proceedings. Edited by W. Schempp and K. Zeller. VI. 290 pages. 1977

Vol. 572: Sparse Matrix Techniques, Copenhagen 1976. Edited by V. A. Barker. V, 184 pages. 1977.

Vol. 573: Group Theory, Canberra 1975. Proceedings. Edited by R. A. Bryce, J. Cossey and M. F. Newman. VII, 146 pages. 1977.

Vol. 574: J. Moldestad, Computations in Higher Types. IV, 203 pages. 1977.

Vol. 575: K-Theory and Operator Algebras, Athens, Georgia 1975. Edited by B. B. Morrel and I. M. Singer. VI, 191 pages. 1977.

Vol. 576: V. S. Varadarajan, Harmonic Analysis on Real Reductive Groups. VI, 521 pages. 1977.

Vol. 577: J. P. May, E_∞ Ring Spaces and E_∞ Ring Spectra. IV, 268 pages. 1977.

Vol. 578: Séminaire Pierre Lelong (Analyse) Année 1975/76. Edité par P. Lelong. VI, 327 pages. 1977.

Vol. 579: Combinatoire et Représentation du Groupe Symétrique, Strasbourg 1976. Proceedings 1976. Edité par D. Foata. IV, 339 pages. 1977.

Vol. 580: C. Castaing and M. Valadier, Convex Analysis and Measurable Multifunctions. VIII, 278 pages. 1977.

Vol. 581: Séminaire de Probabilités XI, Université de Strasbourg. Proceedings 1975/1976. Edité par C. Dellacherie, P. A. Meyer et M. Weil. VI, 574 pages. 1977.

Vol. 582: J. M. G. Fell, Induced Representations and Banach *-Algebraic Bundles. IV, 349 pages. 1977.

Vol. 583: W. Hirsch, C. C. Pugh and M. Shub, Invariant Manifolds. IV, 149 pages. 1977.

Vol. 584: C. Brezinski, Accélération de la Convergence en Analyse Numérique. IV, 313 pages. 1977.

Vol. 585: T. A. Springer, Invariant Theory. VI, 112 pages. 1977.

Vol. 586: Séminaire d'Algèbre Paul Dubreil, Paris 1975–1976 (29ème Année). Edited by M. P. Malliavin. VI, 188 pages. 1977.

Vol. 587: Non-Commutative Harmonic Analysis. Proceedings 1976. Edited by J. Carmona and M. Vergne. IV, 240 pages. 1977.

Vol. 588: P. Molino, Théorie des G-Structures: Le Problème d'Equivalence. VI, 163 pages. 1977.

Vol. 589: Cohomologie l-adique et Fonctions L. Séminaire de Géométrie Algébrique du Bois-Marie 1965–66, SGA 5. Edité par L. Illusie. XII, 484 pages. 1977.

Vol. 590: H. Matsumoto, Analyse Harmonique dans les Systèmes de Tits Bornologiques de Type Affine. IV, 219 pages. 1977.

Vol. 591: G. A. Anderson, Surgery with Coefficients. VIII, 157 pages. 1977.

Vol. 592: D. Voigt, Induzierte Darstellungen in der Theorie der endlichen, algebraischen Gruppen. V, 413 Seiten. 1977.

Vol. 593: K. Barbey and H. König, Abstract Analytic Function Theory and Hardy Algebras. VIII, 260 pages. 1977.

Vol. 594: Singular Perturbations and Boundary Layer Theory, Lyon 1976. Edited by C. M. Brauner, B. Gay, and J. Mathieu. VIII, 539 pages. 1977.

Vol. 595: W. Hazod, Stetige Faltungshalbgruppen von Wahrscheinlichkeitsmaßen und erzeugende Distributionen. XIII, 157 Seiten. 1977.

Vol. 596: K. Deimling, Ordinary Differential Equations in Banach Spaces. VI, 137 pages. 1977.

Vol. 597: Geometry and Topology, Rio de Janeiro, July 1976. Proceedings. Edited by J. Palis and M. do Carmo. VI, 866 pages. 1977.

Vol. 598: J. Hoffmann-Jørgensen, T. M. Liggett et J. Neveu, Ecole d'Eté de Probabilités de Saint-Flour VI – 1976. Edité par P.-L. Hennequin. XII, 447 pages. 1977.

Vol. 599: Complex Analysis, Kentucky 1976. Proceedings. Edited by J. D. Buckholtz and T. J. Suffridge. X, 159 pages. 1977.

Vol. 600: W. Stoll, Value Distribution on Parabolic Spaces. VIII, 216 pages. 1977.

Vol. 601: Modular Functions of one Variable V, Bonn 1976. Proceedings. Edited by J.-P. Serre and D. B. Zagier. VI, 294 pages. 1977.

Vol. 602: J. P. Brezin, Harmonic Analysis on Compact Solvmanifolds. VIII, 179 pages. 1977.

Vol. 603: B. Moishezon, Complex Surfaces and Connected Sums of Complex Projective Planes. IV, 234 pages. 1977.

Vol. 604: Banach Spaces of Analytic Functions, Kent, Ohio 1976. Proceedings. Edited by J. Baker, C. Cleaver and Joseph Diestel. VI, 141 pages. 1977.

Vol. 605: Sario et al., Classification Theory of Riemannian Manifolds. XX, 498 pages. 1977.

Vol. 606: Mathematical Aspects of Finite Element Methods. Proceedings 1975. Edited by I. Galligani and E. Magenes. VI, 362 pages. 1977.

Vol. 607: M. Métivier, Reelle und Vektorwertige Quasimartingale und die Theorie der Stochastischen Integration. X, 310 Seiten. 1977.

Vol. 608: Bigard et al., Groupes et Anneaux Réticulés. XIV, 334 pages. 1977.

Vol. 609: General Topology and Its Relations to Modern Analysis and Algebra IV. Proceedings 1976. Edited by J. Novák. XVIII, 225 pages. 1977.

Vol. 610: G. Jensen, Higher Order Contact of Submanifolds of Homogeneous Spaces. XII, 154 pages. 1977.

Vol. 611: M. Makkai and G. E. Reyes, First Order Categorical Logic. VIII, 301 pages. 1977.

Vol. 612: E. M. Kleinberg, Infinitary Combinatorics and the Axiom of Determinateness. VIII, 150 pages. 1977.

Vol. 613: E. Behrends et al., L^p-Structure in Real Banach Spaces. X, 108 pages. 1977.

Vol. 614: H. Yanagihara, Theory of Hopf Algebras Attached to Group Schemes. VIII, 308 pages. 1977.

Vol. 615: Turbulence Seminar, Proceedings 1976/77. Edited by P. Bernard and T. Ratiu. VI, 155 pages. 1977.

Vol. 616: Abelian Group Theory, 2nd New Mexico State University Conference, 1976. Proceedings. Edited by D. Arnold, R. Hunter and E. Walker. X, 423 pages. 1977.

Vol. 617: K. J. Devlin, The Axiom of Constructibility: A Guide for the Mathematician. VIII, 96 pages. 1977.

Vol. 618: I. I. Hirschman, Jr. and D. E. Hughes, Extreme Eigen Values of Toeplitz Operators. VI, 145 pages. 1977.

Vol. 619: Set Theory and Hierarchy Theory V, Bierutowice 1976. Edited by A. Lachlan, M. Srebrny, and A. Zarach. VIII, 358 pages. 1977.

Vol. 620: H. Popp, Moduli Theory and Classification Theory of Algebraic Varieties. VIII, 189 pages. 1977.

Vol. 621: Kauffman et al., The Deficiency Index Problem. VI, 112 pages. 1977.

Vol. 622: Combinatorial Mathematics V, Melbourne 1976. Proceedings. Edited by C. Little. VIII, 213 pages. 1977.

Vol. 623: I. Erdelyi and R. Lange, Spectral Decompositions on Banach Spaces. VIII, 122 pages. 1977.

Vol. 624: Y. Guivarc'h et al., Marches Aléatoires sur les Groupes de Lie. VIII, 292 pages. 1977.

Vol. 625: J. P. Alexander et al., Odd Order Group Actions and Wit Classification of Innerproducts. IV, 202 pages. 1977.

Vol. 626: Number Theory Day, New York 1976. Proceedings. Edited by M. B. Nathanson. VI, 241 pages. 1977.

Vol. 627: Modular Functions of One Variable VI, Bonn 1976. Proceedings. Edited by J.-P. Serre and D. B. Zagier. VI, 339 pages. 1977

Vol. 628: H. J. Baues, Obstruction Theory on the Homotopy Classification of Maps. XII, 387 pages. 1977.

Vol. 629: W.A. Coppel, Dichotomies in Stability Theory. VI, 98 pages 1978.

Vol. 630: Numerical Analysis, Proceedings, Biennial Conference Dundee 1977. Edited by G. A. Watson. XII, 199 pages. 1978.

Vol. 631: Numerical Treatment of Differential Equations. Proceedings 1976. Edited by R. Bulirsch, R. D. Grigorieff, and J. Schröder. X 219 pages. 1978.

Vol. 632: J.-F. Boutot, Schéma de Picard Local. X, 165 pages. 1978

Vol. 633: N. R. Coleff and M. E. Herrera, Les Courants Résiduels Associés à une Forme Méromorphe. X, 211 pages. 1978.

Vol. 634: H. Kurke et al., Die Approximationseigenschaft lokaler Ringe IV, 204 Seiten. 1978.

Vol. 635: T. Y. Lam, Serre's Conjecture. XVI, 227 pages. 1978

Vol. 636: Journées de Statistique des Processus Stochastiques, Grenoble 1977, Proceedings. Edité par Didier Dacunha-Castelle et Bernard Van Cutsem. VII, 202 pages. 1978.

Vol. 637: W. B. Jurkat, Meromorphe Differentialgleichungen. V. 194 Seiten. 1978.

Vol. 638: P. Shanahan, The Atiyah-Singer Index Theorem, An Introduction. V, 224 pages. 1978.

Vol. 639: N. Adasch et al., Topological Vector Spaces. V, 125 pages 1978.